Wolfgang Felzmann

Contributions to the Total Synthesis of Branimycin

Wolfgang Felzmann

Contributions to the Total Synthesis of Branimycin

Südwestdeutscher Verlag für Hochschulschriften

Imprint
Any brand names and product names mentioned in this book are subject to trademark, brand or patent protection and are trademarks or registered trademarks of their respective holders. The use of brand names, product names, common names, trade names, product descriptions etc. even without a particular marking in this work is in no way to be construed to mean that such names may be regarded as unrestricted in respect of trademark and brand protection legislation and could thus be used by anyone.

Cover image: www.ingimage.com

Publisher:
Südwestdeutscher Verlag für Hochschulschriften
is a trademark of
International Book Market Service Ltd., member of OmniScriptum Publishing Group
17 Meldrum Street, Beau Bassin 71504, Mauritius

Printed at: see last page
ISBN: 978-3-8381-5099-4

Zugl. / Approved by: Wien, Universität Wien, Diss., 2007

Copyright © Wolfgang Felzmann
Copyright © 2015 International Book Market Service Ltd., member of OmniScriptum Publishing Group
All rights reserved. Beau Bassin 2015

TABLE OF CONTENTS

Table of Contents 1
Background and Introduction 4
1   Antibiotics 4
    1.1   The plight of past centuries removed 4
    1.2   Modes of action 8
    1.3   Antibiotic resistance – a fight for a lost cause? 9
    1.4   Development of new antibiotic drugs 11
        1.4.1   Finding new leads 11
        1.4.2   Sources of lead structures 11
    1.5   Branimycin 12
2   Branimycin and related Natural Products 14
    2.1   The Nargenicins 14
    2.2   Coloradocin / Luminamicin / Lustromycin 18
    2.3   Biosynthesis 19
    2.4   Comparison and conclusion 24
3   Previous Synthetic Work 26
    3.1   Approaches towards the nargenicin family 26
        3.1.1   Kallmerten's synthesis of 18-desoxynargenicin A_1 26
        3.1.2   Kallmerten's approach towards 18-oxygenated nargenicins 29
        3.1.3   Jones' approach to the oxo-bridged *cis*-octalin 31
        3.1.4   Roush' IMDA approach to the *cis*-octalin core 32
        3.1.5   Roush' biomimetic approach towards nargenicin A_1 34
        3.1.6   Gössinger's approach towards nodusmycin 36
    3.2   Approaches towards Coloradocin 38
    3.3   Other approaches towards branimycin in the Mulzer group 39
        3.3.1   Quinic acid based INOC approach 39
        3.3.2   Lactone-templated INOC Approach 40
        3.3.3   Double-Claisen RCM Approach 41
4   Outline of the Synthetic Plan 43

Results and Discussion 46
5   1st Generation Approach 48
6   2nd Generation Approach 55
    6.1   Concept 55
    6.2   Chiral tether synthesis 57
    6.3   Efforts towards a C-2 functionalization 61
    6.4   *E,Z*-Diene building block 66
        6.4.1   Wittig-approach 66
        6.4.2   Cross-coupling approach 69
        6.4.3   Cross-coupling / lactol approach 73
7   3rd Generation Approach 80
    7.1   Concept 80
    7.2   Approaches towards the *Z,E,Z,E*-macrolactone 85
        7.2.1   C-5 – C-6 Stille coupling approach 88
        7.2.2   'Stitching' coupling approach 92
        7.2.3   C-7 – C-8 Stille coupling approach 95
        7.2.4   Alternative C-7 – C-8 couplings 102
        7.2.5   Grob-type ring expansion strategy 111
    7.3   Optimisation of the synthesis of macrolactone 224 115
        7.3.1   Alternative Synthesis of the *Z*-C-4 – C-5 double bond 116
        7.3.2   Reliable C-11 – C-10 Olefination 119
        7.3.3   Optimisation of the C-7 – C-8 intramolecular Stille-coupling 125
    7.4   *cis*-Hexalin functionalisation 129
8   Synthesis of the C-13 – C-18 Side Chain 138
    8.1   Concept 138
    8.2   Prior work 139
    8.3   Diastereoselective propargylation approach 142
9   Conclusion and Outlook 155
    9.1   *Cis*-octalin core fragment 155
    9.2   C-13 – C-18 Side Chain 157
    9.3   Outlook / Possible improvements 159
    9.4   Proposed Endgame 161

Experimental Section 162
10 General 162
    10.1 Solvent Purification 162
    10.2 Reaction Control 162
    10.3 Column Chromatography 162
    10.4 NMR-Spectroscopy 163
    10.5 Other Spectroscopic Methods 163
    10.6 X-Ray Analysis 163
11 Experimental Procedures 165
    11.1 First Generation Approach 165
    11.2 Second Generation Approach 174
    11.3 Third Generation Approach 206
    11.4 Side Chain Synthesis 244
12 Appendix 263
    12.1 Used Abbreviations 263
    12.2 Single-crystal diffraction data 265
    12.3 Literature 269
Graphical Abstract 277
Abstract 278
Zusammenfassung 279

BACKGROUND AND INTRODUCTION

1 ANTIBIOTICS

1.1 The plight of past centuries removed

Since the introduction of the first antibiotics to the market in 1933, the mortality rate for infectious diseases has sunk from about 35 % in 1910 to approximately 4 % in 1990 (also because of higher hygienic standards and better nutrition).[1] These numbers are a good foundation for the argument that the invention and application of antibiotics represents one of the major milestones of medicinal history and is for sure one of the major scientific achievements of the last century.[2]

The first antibiotics applicable in a broad fashion were discovered in the 1930's. The first synthetic antibiotic prontosil (**1**) was discovered in 1935 by the physician G. Domagk at Bayer AG – this compound, originally developed as an azo dye, turned out to show strong antibiotic activity. (Figure 1) Due to the simple structure, industrial production of this sulfonamide azo compound was easily possible, which found widespread use during the Second World War to prevent wound infections. The up to date most commonly known antibiotic, penicillin was discovered by A. Fleming in 1929.[3] However, it took more than ten years before the derivative penicillin G (**2**) was first used in humans, when H. W. Florey and E. B. Chain further investigated the findings of Fleming. War demands then boosted the development and led to industrial production only 3 years later in 1943. Inspired by these success story, the screenings of different soil organisms led to the isolation of further antibiotics. Cephalosporin, another β-lactam antibiotic was isolated in 1945 – many derivatives of this substance

---

[1] Walsh, C. T.; Wright, G. *Chem. Rev.* **2005**, *105*, 391 – 393.

[2] Nussbaum, F.; Brands, M.; Hinzen, B.; Weigand, S.; Häbich, D. *Angew. Chem. Int. Ed.* **2006**, *45*, 5072 – 5129; *Angew. Chem.* **2006**, *118*, 5194 – 5254.

[3] Fleming, A. *Br. J. Exp. Pathol.* **1929**, *10*, 226–236.

(e.g. cephazolin **4**) were later marketed for both parenteral and oral application). Chloramphenicol (**3**) is a bacteriostatic that was first isolated in 1947 from *Streptomyces venezuelae* – it was first introduced into clinical practice in 1949.

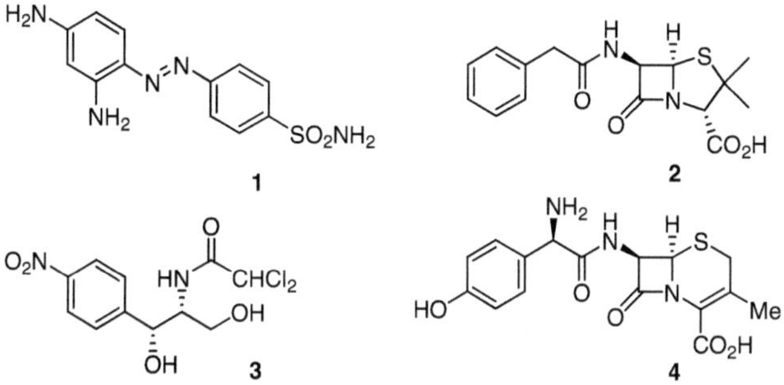

Figure 1: Prontosil (**1**), penicillin G (**2**), chloramphenicol (**3**) and cephazolin (**4**)

The β-lactams were regarded as fairly complex structures at this time (the molecular structure of penicillin was unknown when it was developed as a drug – R. B. Woodward spent some time during the Second World War synthesizing the 'wrong' structure of penicillin[4]).

However, even far more complicated antibiotics were isolated (although again, without elucidation of their structure). (Figure 2) Streptomycin (**6**) was isolated from a stem of *Streptomyces* in the group of A. Waksman in 1943. It was the first antibiotic to be effective against tuberculosis.

Tetracycline (**5**) is one member of large group of natural products, all signified by four condensed hydrocarbon rings originating from a polyketide biosynthetic pathway. It is derived by reduction from the natural product oxytetracycline, which was isolated from the soil bacterium *Streptomyces rimosus*.

Erythronolide (**7**) was isolated from *Saccharopolyspora erythraea*, found in soil samples from the Philippines. Structurally, it is a 14-membered macrolactone ring

---

[4] Originally, other structures than the correct β-lactam had been put forward. For a review on the syntheses of misassigned natural products, see: Nicolaou, K. C.; Snyder, S. A. *Angew. Chem. Int. Ed.* **2005**, *44*, 1012-1044; *Angew. Chem.* **2005**, *117*, 1036 – 1069.

bearing ten asymmetric centers, 2 of them being glycosidated. Its complex structure was regarded as too complex to be accessed by chemical synthesis, which was disproved 30 years later by the first total synthesis by Woodward in 1981.[5]

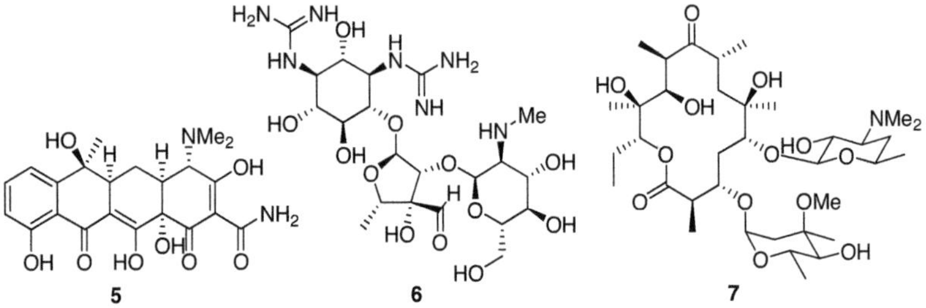

Figure 2: Tetracycline (**5**), streptomycin (**6**) and erythromycin (**7**)

Vancomycin (**8**), a non-ribosomal glycopeptide, was isolated from the actinobacterium *Amycolatopsis orientali*, which was found in a soil sample collected in the jungle of Borneo. (Figure 3) After approval for human use in 1958, it was initially used for the treatment of penicillin-resistant *Staphylococcus aureus* – however, it was used only reluctantly because impurities in the initial formulation led to inherent toxicity problems. 20 years later, cleaner formulations could be produced that did not show these problematic side-effects. From this time on, vancomycin (**8**) was regarded as a very reliable last-resort antibiotic, with basically no bacterial resistances developed.[6] Due to the impressive complexity of this molecule the first successful total synthesis was achieved only in 1998.[7]

---

[5] a) Woodward, R. B. *et al*. *J. Am. Chem. Soc*. **1981**, *103*, 3210 – 3213; b) Woodward, R. B. *et al*. *J. Am. Chem. Soc*. **1981**, *103*, 3213 – 3215; c) Woodward, R. B. *et al*. *J. Am. Chem. Soc*. **1981**, *103*, 3215 – 3217.

[6] Moellering, R. C. Jr. *Clin. Infect. Dis*. **2006**, *42*, S3–S4.

[7] Nicolaou, K. C.; Mitchell, H. J.; Jain, N. F.; Winssinger, N.; Hughes, R.; Bando, T. *Angew. Chem. Int. Ed*. **1999**, *38*, 240 – 244; *Angew. Chem*. **1999**, *111*, 253 – 255.

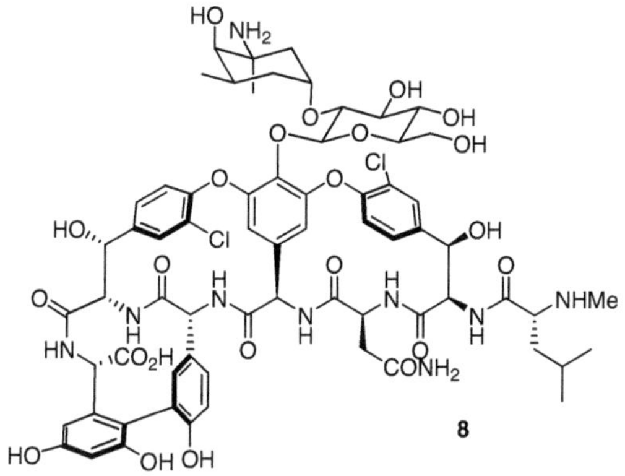

Figure 3: Vancomycin (**8**)

The rifamycins were first isolated in 1957 from soil bacteria found in the south of France named *Amycolatopsis mediterranei*. Both the natural product rifamycin B, as well as several derivatives, like rifampicin (**10**), were marketed. (Figure 4) This class is widely used for the treatment of tuberculosis and leprosy.

In the 1960's also the first fully synthetic bactericidal drug was developed – nalidixic acid. It was the first of a whole series of quinolones, which were later improved by fluorination of the 6-position. Ciprofloxacine (**9**) is one of most widely known representatives of the group of fluoroquinolones.

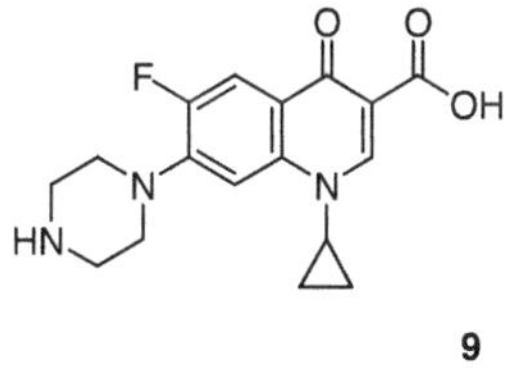

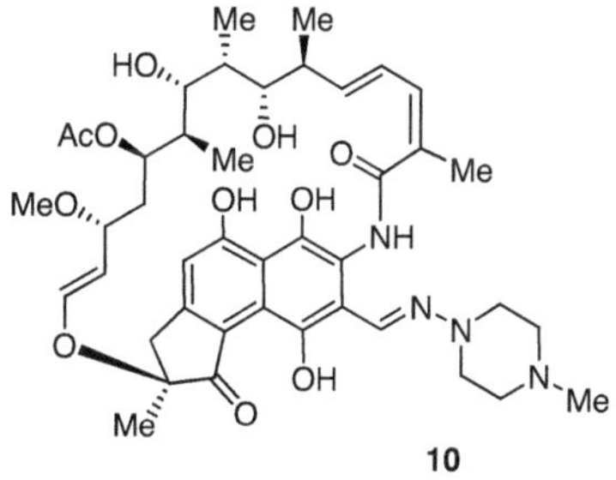

Figure 4: Ciprofloxacin (**9**) and rifampicin (**10**)

From the 1970's to the late 1990's, no new antibiotically active substance classes were introduced. This only changed in 2000, when the oxazolidinones, a new class of synthetically developed antibiotics were introduced, with linezolid (**11**) being the first drug marketed from this class. (Figure 5) Furthermore, in 2005 the lipopeptide daptomycin (**12**), which was isolated from the soil saprotroph *Streptomyces roseosporus,* was introduced to the market under the name CUBICIN by Cubist Pharmaceuticals in the United States.

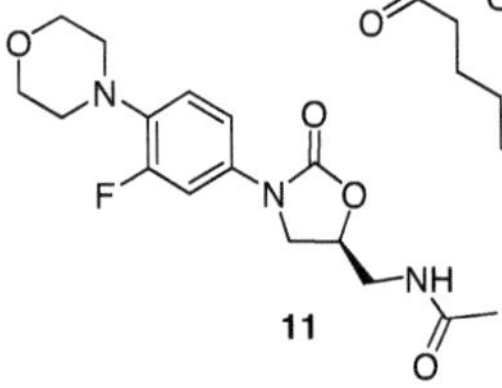

Figure 5: Linezolid (**11**) and daptomycin (**12**)

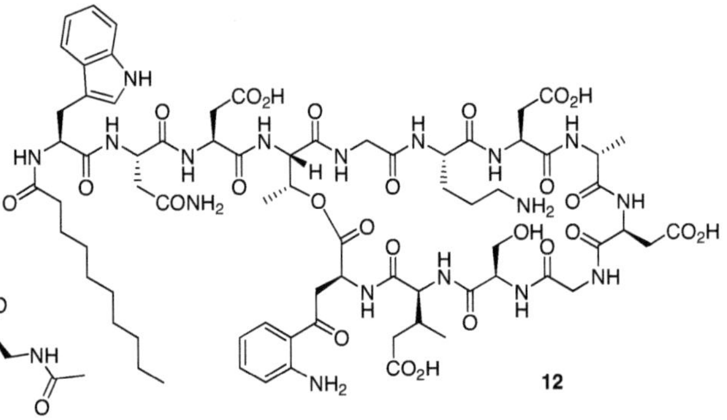

1.2 Modes of action

Antibiotics can be divided into two large groups: bacteriostatic agents inhibit the growth and cell division, while bactericidal drugs actually kill the bacteria. Besides this classification, 4 main mechanisms of action can be differentiated:

- Inhibition of folate synthesis: This mode of action is followed by the sulfonamides, which are active-site inhibitors for the folate synthase.
- Inhibition of cell wall growth: All β-lactam antibiotics belong to this group – they inhibit the alanin-transcriptase, which in turn hinders bacterial cell wall growth. Vancomycin (**8**) exhibits a different mode of action. It damages bacterial cell walls by inhibiting the incorporation of *N*-acetylmuramic acid (NAM) – and *N*-acetylglucosamine (NAG)-peptide subunits into the peptidoglycan matrix, which is a major constituent of Gram-positive cell walls. The lipopeptide daptomycin (**12**) disrupts the cell membrane, leading to depolarisation of the cell wall.
- Protein biosynthesis: Several antibiotics interfere with ribosomal protein synthesis – as this is a quite complex process, different steps in this sequence

can be targeted by antibiotics. The aminoglycosides target the 30S subunit of prokaryotic 70S ribosomes, inhibiting the binding of the aminoacyl-tRNA to its acceptor position. This in turn blocks the translation of the genes and hampers growth. The mode of action of chloramphenicol (**3**) and linezolid (**11**) is similar – both block the transpeptidation. With erythromycin, the binding site differs: it binds to the 23S rRNA molecule in the 50S subunit of the bacterial ribosome, blocking the exit of the growing peptide chain thus inhibiting the translocation of peptides. Tetracycline (**5**) binds to the 16S part of the 30S ribosomal subunit, competing with tRNA for A-site binding. Also in this case there is a good selectivity for prokaryotic ribosomes, resulting in a low toxicity for humans.

- <u>DNA/RNA synthesis:</u> The fluoroquinolones inhibit the bacterial DNA gyrase, the enzyme responsible in prokaryotic cells for the coiling of the DNA-strands. This hinders DNA replication and transcription, leading to cell death. Rifampicin (**10**) obstructs the DNA-dependent RNA polymerase by binding to its β-subunit. This prevents the transcription of mRNA.

1.3 Antibiotic resistance – a fight for a lost cause?

The development of antibiotic resistance is a fundamentally natural process. The presence of antibiotics exerts an evolutionary stress on a microbial population to mutate and in turn select possible resistant organisms. The observed resistance is the consequence of the application of one of the following mechanisms by the bacteria: The targeted protein can be altered, which reduces the affinity of the antibiotic to the receptor site, or the cell metabolism is changed to reduce the uptake of the antibiotic. Both strategies can be regarded as 'passive defences'. On the other hand, the active defences of a bacterial organism are the enzymatic deactivation of antibiotics, as well as an increased cell efflux. Such phenomena were known for a long time, but certain antibiotics seemed quite robust and no signs of antibiotic resistance were observed. Furthermore, the impressive number of different antibiotics, followed by the development of multiple generations of drugs based on the initial structures gave the impression, that there was no need for the development of new antibiotics.[2] In fact, bacterial infections were thought to be soon extinct in the early 1970's, which is nicely illustrated by the quotation attributed to the U.S. surgeon General W. H. Stewart that "the book on infectious diseases could now be closed". With the need for new antibiotics getting questioned by reknown scientists, industry and public interest started to neglect the field of antibiotic research, which resulted in an innovation gap of at least 2 decades. In this time period, several incidents led to a revision of this assumption. For a long time, multiresistant stems of *Staphylococcus aureus* (termed 'methicillin resistant *S. aureus*' – MRSA) could still be adressed successfully with vancomycin.[8] However, in the late 1980's, a decreased response of *enterococci* towards vancomycin was first noticed. With the discovery of vancomycin-resistant *Staphylococcus aureus* (VRSA)[9] in 2002, the situation became more critical – especially in hospitals, where the number of multidrug-resistant stems encountered in intensive-care units has dramatically increased over the last 15 years. Up to date, in most of these cases multidrug-therapies can still be applied successfully. However, it depends on the success of such treatments, if a patient dies or lives. Due to the diminishing susceptibility of bacteria towards antibiotics and, as treatment options for certain microorganisms become increasingly scarce, experts warn of a 'shadow' epidemic,[8] which might ultimately catapult physicians back into pre-antibiotic times.

Nowadays, infectious diseases are still the second major cause of death worldwide and the third leading cause of death in developed countries. This situation can only be kept in check (a turning point seems unlikely in the situation today) with persistent discovery and development of new drugs.

---

[8] Levy, S. B. *Adv. Drug Deliv. Rev.* **2005**, *57*, 1446 – 1450.

[9] a) Hiramatsu, K.; Okuma, K.; Ma, X. X.; Yamamoto, M.; Hori, S.; Kapi, M. *Curr. Opin. Infect. Dis.* **2002**, *15*, 407 – 413; b) Weigel, L. M.; Clewell, D. B.; Gill, S. R.; Clark, N. C.; McGougal, L. K.; Flannagan, S. E.; Kolonay, J. F.; Shetty, J.; Killgore, G. E.; Tenover, F. C. *Science* **2003**, *302*, 1569 – 1571.

1.4 Development of new antibiotic drugs

1.4.1 Finding new leads

Antibiotics are one group of pharmaceuticals which did not profit from the introduction of high-throughput screening (HTS) methods. The credo "one target, one drug" cannot be applied to antibiotics with the same ease, as the mode of action of several antibiotics is multifaceted and cannot be reduced to the simple interaction of the drug with one single target. Indeed, there is no antibiotic on the market stemming from an HTS-procedure. Instead, the more classical approach, using minimum inhibitory concentration (MIC) tests on various bacterial cultures, is still used and has proven to be an indispensable source of information for the validation of potential lead structures. After determination of the 'macroscopic' activity, the mode of action of the antibiotic has to be determined. This is to find out if possible cross-resistances from other antibiotics have to be expected, in the case both are addressing the same molecular targets.

1.4.2 Sources of lead structures

Given the grand interest in the identification of new lead structures, several ways to obtain such are followed: Using the combination of state-of-the-art biochemical methodology (reversed genomics) and powerful molecular modelling techniques, the binding motives of natural products can be identified. Aided by *in-silico* studies of the binding site, simpler synthetic molecules can be conceived. These can then serve as scaffolds for combinatorial and parallel synthesis.

In the course of the antibiotic research in the last 50 years, many antibiotic natural products have been identified and patented, but were never exploited for several reasons. This dormant arsenal of antibiotic treasures can be thoroughly explored using modern SAR studies, which, combined with intelligent functionalisation/degradation/modification, can then lead to new drug candidates.

Finally, the screening of natural sources still remains one of the most attractive ways to find new drug targets. Even though fully synthetic drugs had a tremendous success story in many fields of pharmaceutical research, this was not the case in the field of antibiotic development, as can be easily seen by inspection of chapter 1.1. Using natural products as scaffolds for antibiotics is certainly a key to access bacterial vulnerability.

To ascertain the discovery of 'new' natural products which have not been described before, the screening strategies of natural sources have to be adopted or modified considering the following points:

- Application of different cultivation techniques
- Investigation of totally different organisms
- Genetic modification of the organisms
- Intelligent testing procedures
- Accelerated structure determination

1.5 Branimycin

In 1998, the Laatsch group at the University of Göttingen found, upon screening of the culture broth of the *Streptomyces* stem GW 60/1571,[10] it to be active against *E. coli*, *Bacillus subtilis*, *Streptomyces viridochromogenes* and *Staphylococcus aureus*. Fractioning and purification showed this activity to be derived from a substance, whose NMR signals showed no match with published structures in the databases. The basic C-C connectivities could be determined by extensive NMR analysis – full structural assignment was possible by comparison with the NMR data available for a related compound nargenicin A_1 (**15**) (*vide infra*), followed by adaption of the stereochemistry based on NOE-interactions. The substance was named branimycin (**14**).

---

[10] Speitling, M. *Dissertation*, Universität Göttingen, **1998**.

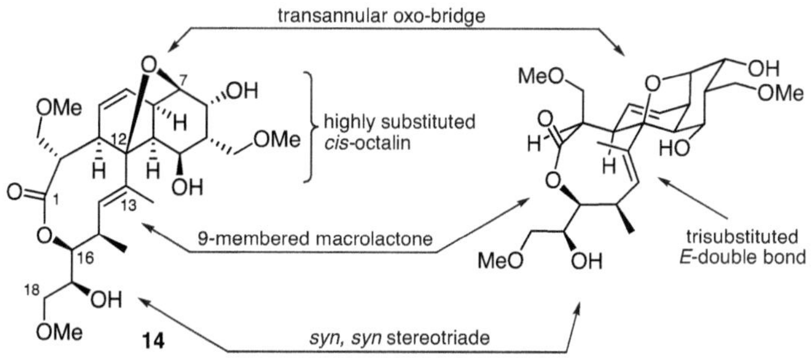

Figure 6: Assigned structure of branimycin (**14**)

The structure of branimycin (**14**) is characterised by a densely functionalised *cis*-octalin core, where C-7 and C-12 are spanned by a transannular oxo-bridge. Annulated to the octalin is a 9-membered macrolactone ring, containing a trisubstituted *E*-double bond. The stereocenters on the C-13 – C-18 polyketide chain of the macrolactone are in a *syn,syn* – relationship.

Due to the interesting biological activity of branimycin, besides its compelling structure, it was chosen as a total synthesis target in the Mulzer group. Such total syntheses are important in a two-fold sense:
First, it is the only way to proof the structure proposed on the basis of spectroscopic analyses – in the case of branimycin, the main point of concern is the C-17 stereochemistry, which is inverted in comparison to related natural products. Its complex structure is, even with the aid of modern synthetic methods, still a venerable challenge. Its synthesis will certainly show the applicability of known methods on such a complex target, and inspire the development of new ones as well.
Second, a total synthesis serves as an excellent basis to get comprehensive access to all parts of the molecule, which facilitates later SAR studies.

The aim of this thesis was to develop an efficient synthetic strategy to access this complex molecule or a major building block thereof.

2   BRANIMYCIN AND RELATED NATURAL PRODUCTS

As mentioned before, branimycin (**14**) belongs to the class of the nargenicin antibiotics – its structure determination would not have been so easily possible without the prior structural assignment of the parent compound nargenicin A_1 (**15**). The structural diversity of this class of compounds, the history of its discovery and also the available information about the biosynthesis are summarised in this chapter.

2.1   The Nargenicins

In 1977, scientists from Pfizer noted that a newly discovered stem of *Nocardia*, named *Nocardia argentinensis* nov. gen. (ATCC 31306), produced novel antibiotics upon aerobic submerged fermentation.[11] One of these antibiotics, CP 47,777, and its isolation procedure, was subsequently patented, although without the disclosure of its structure. Some years later, the structure of CP 47,777 was published, making it the first member of a new class of antibiotics, and the name nargenicin A_1 (**15**) was proposed.[12]

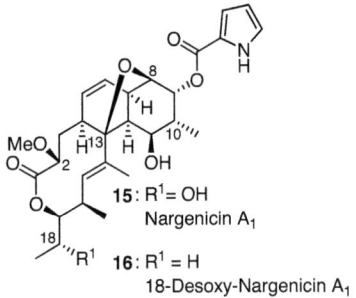

Figure 7: Nargenicin A_1 and its 18-Deoxy-derivative

---

[11] Celmer, W. D.; Cullen, W. P.; Moppett, C. E.; Jefferson, M. T.; Huang, L. H.; Shibakawa, R.; Tone, J. *United States Patent* US 4,418,883, **1979**.

[12] Celmer, W. D.; Chmurny, G. N.; Moppett, C. E.; Ware, R. S.; Watts, P. C.; Whipple, E. B. *J. Am. Chem. Soc.* **1981**, *102*, 4203 – 4209.

The structure elucidation was based on extensive NMR-analysis and allowed the authors to assign the relative stereochemistry on the rigid *cis*-octalin ring; however, the stereochemistry at the position 2, 16, 17, 18 on the macrolactone ring could not be assigned. The structure depicted in Figure 7 is a result of further research on this matter (*vide infra*). Degradation studies showed that the C-18 hydroxy function could be acylated selectively; basic ethanolysis resulted in formation of both the ring-opened ethyl ester and, as a side product, the ring-expanded macrolactone to the C-18 oxygen. In later biosynthetic studies[13] this ring-expanded Nargenicin A_1, named isonargenicin was found to be always present as a minor component in the fermentation of **15**.

Some time later it could be shown, that 18-desoxynargenicin (**16**), another minor constituent in these fermentations, can be synthesised by selective deoxygenation of Nargenicin A_1 (**15**),[14] thus indirectly proving this specific structural feature.

Shortly after the publication of the structure of nargenicin A_1 (**15**), scientists from Upjohn reported that the soil organism *Saccharopolyspora hirsuta* produced upon fermentation a complex mixture of antibiotics, out of which nodusmycin (**17**) was isolated.[15] It corresponds to nargenicin A_1 lacking the 2-pyrrol-carboxylate in the 9 position.

In contrast to nargenicin, nodusmycin (**17**) could be crystallised and analysed by X-ray analysis, thus proving for the first time the relative configuration of both the *cis*-octalin and the polyketide macrolactone.

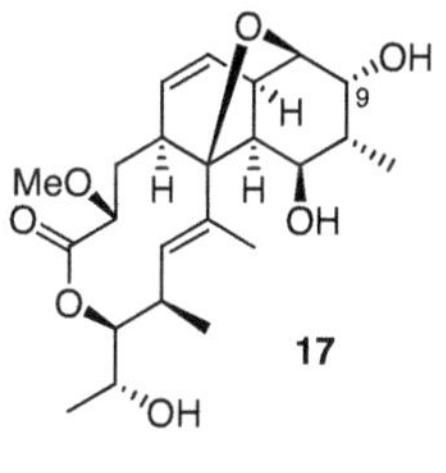

Figure 8: Nodusmycin

---

[13] Cane, D. E.; Yang, C.-C. *J. Am. Chem. Soc.* **1984**, *106*, 784 – 787.
[14] Magerlein, B. J.; Reid, R. J. *J. Antibiot.* **1982**, *35*, 254 – 255.
[15] Whaley, H. A.; Chidester, C. G.; Mizsak, S. A.; Wnuk, R. J. *Tetrahedron Lett.* **1980**, *21*, 3659 – 3662.

At the same time, the Pfizer group patented the isolation of the antibiotics CP 51,467, CP 52,726 and CP 52,748 from the species *Nocardia argentinenis* Huang ATCC 31438,[16] again without disclosure of its structure. However, four years the isolation of another antibiotic from the nargenicin family was patented by the same group: nargenicin C_1 (**21**)[17] had been obtained from a new stem of *Nocardia*, isolated from a soil sample collected in Georgia, U.S.A (ATCC 39177). In this patent, along with the structure of **21**, also the structures of nargenicins B_1-B_3 (**18-20**) were published, which may be attributed to the previously patented substances CP 51,467, CP 52,726 and CP 52,748.

18: R^2 = OMe R^3 = OMe R^4 = OH
Nargenicin B_1

19: R^2 = OMe R^3 = OH R^4 = H
Nargenicin B_2

20: R^2 = H R^3 = OMe R^4 = OH
Nargenicin B_3

21: Nargenicin C_1

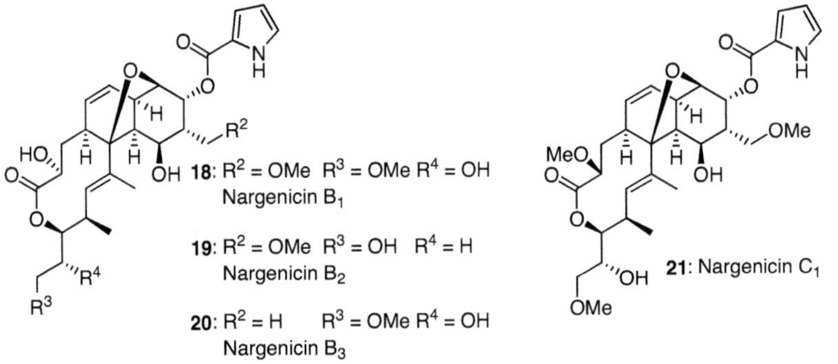

Figure 9: Nargenicins B_{1-3} and C_1

In general, the nargenicin-type antibiotics can be described by the oxygen-bridged *cis*-octalin system, annulated to the 10-membered macrolactone ring. The existence of a 2-pyrrol-carboxylate ester at C-9 seems to be a common feature only for the nargenicins, but is missing in other members of this family (e.g. nodusmycin, coloracidin – *vide*

---

[16] Celmer, W. D.; Cullen, W. P.; Moppett, C. E.; Jefferson, M. T.; Huang, L. H.; Shibakawa, R.; Tone, J. *United States Patent* US 4,224,314, **1980**.

[17] Celmer, W. D.; Cullen, L. H.; Shibakawa, R.; Tone, J. *United States Patent* US 4,436,747, **1984**.

infra).[18] The oxygenation at the C-10-methyl group, at C-18 and C19 varies, as well as the configuration of the C-2 oxygen. (Figure 10)

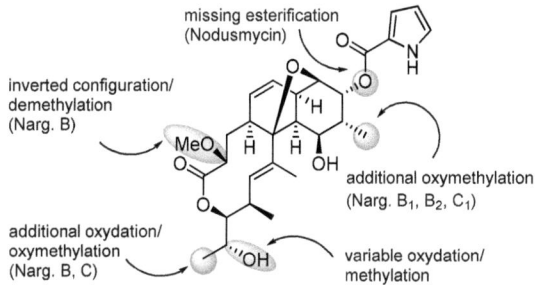

Figure 10: Variations in the nargenicin family

All members of the Nargenicin family show distinct antibiotic activity, the most against *Staphylococcus aureus*, including strains that are resistant to other antibiotics. They are further active against *Bacillus subtilis*, *Streptococcus pyogenes*, *Pasteurella multocida* and *Neisseria sicca* (Nargenicin A_1) and *Staphylococcus epidermidis* (CP 51,467). These antibacterial activites come along with a significant oral availability (0.5g to 1g as a daily dose), compared to an equally effective parenteral administration of 0.1 to 0.5 g.

In comparison with branimycin (**14**), the largest congruence found with a published member of the nargenicin family is with nargenicin B_1 (**18**). (Figure 11) Nargenicin B_1 also carries methoxy-groups at the C-10 and C-19 methyl groups (nargenicin numbering). In contrast to nargenicin B_1, carbon C-2 is shifted to an exocyclic position in branimycin, resulting in a one-carbon contracted 9-membered macrolactone ring. However, as no NMR data is published for nargenicin B_1, no direct comparison of these two compounds is possible.

---

[18] This might be an arbitrary feature, as this UV-active structural element was used for the detection of nargenicin A_1. In ref. 12, the authors state "The structure suggests a signet ring with a macrolide band, a rigid cage for the stone, and a pyrrole ring for the seal which has so far served to identify additional members of the class."

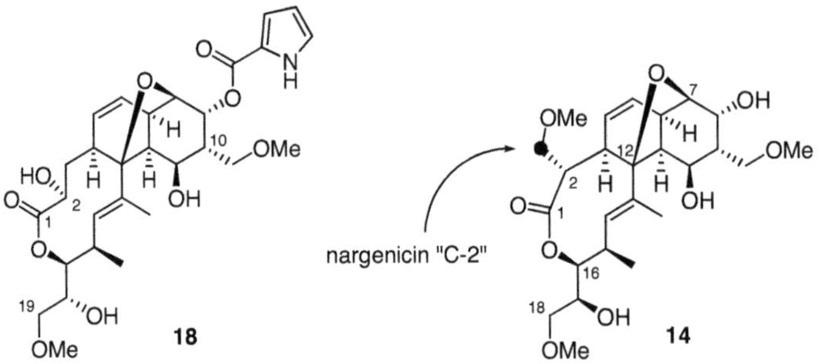

Figure 11: Comparison of nargenicin B₁ (**18**) and branimycin (**14**)

2.2   Coloradocin / Luminamicin / Lustromycin

In 1985, Japanese scientists isolated a new antibiotic from a stem of the genus *Nocardioides*, which they named luminamicin, although without disclosing its structure.[19] Independently, researchers from the Abbott Laboratories isolated the antibiotic coloradocin (**22**) from the organism *Actinoplanes coloradoensis*,[20] which proved identical to luminamicin.[21] (Figure 12) In Coloradocin (**22**) the oxygen bridge connects C-9 – C-13, differing from the nargenicins.

Later, the absolute configuration of the C-11 and C-18 – hydroxy groups were determined by Mosher ester derivatization, combined with molecular dynamics studies.[22] Thus, based on the relative stereochemistry derived from the NMR analysis, the structure depicted in Figure 12 could be assigned. The structure of lustromycin

---

[19] Ōmura, S.; Iwata, R.; Iwai, Y.; Taga, S.; Tanake, Y.; Tomoda, H. *J. Antibiot.* **1985**, *38*, 1322 – 1326.

[20] Jackson, M.; Karwowski, J. P.; Theriault, R. J.; Fernandes, P. B.; Semon, R. C. *J. Antibiot.* **1987**, *40*, 1375 – 1381.

[21] Rasmussen, R. R.; Scherr, M. H.; Whittern, D. N.; Buko, A. M.; McAlpine, J. B. *J. Antibiot.* **1987**, *40*, 1383 – 1393.

[22] Gouda, H.; Sunazuka, T.; Ui, H.; Handa, M.; Sakoh, Y.; Iwai, Y.; Hirono, S.; Ōmura, S. *Proc. Natl. Acad. Sci. U.S.A.* **2005**, *102*, 18286 – 18291.

(23),[23] a closely related antibiotic was determined by NMR-analysis.[24] It should be noted that the stereochemistry at C-2 and C-3 could not be determined, or the required NOE-studies necessary for their assignment were excluded deliberately in the publication.

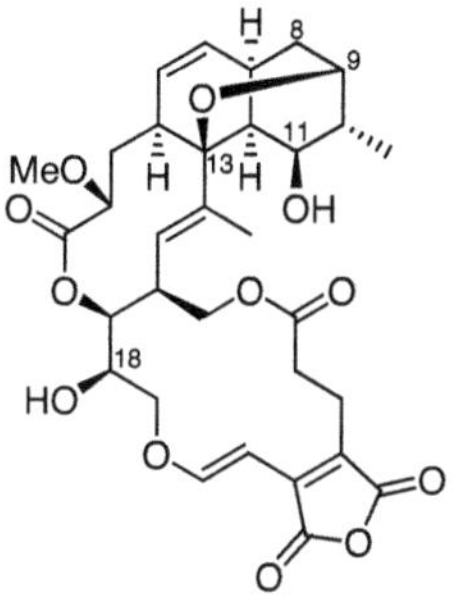

Figure 12: Coloradocin (**22**) and Lustromycin (**23**)

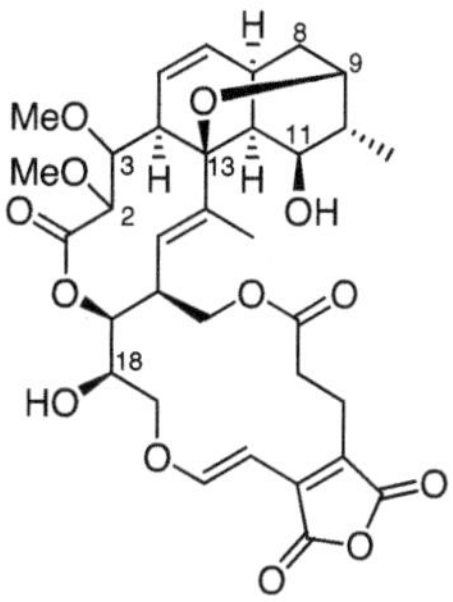

2.3 Biosynthesis

Nargenicin A_1 was soon suspected to be biochemically derived from a polyketide pathway similar to that of other macrolide antibiotics (*e.g.* erythromycin). However, the biosynthetic formation of carbacyclic polyketides was much less understood than that of polyketide macrolactones, and the underlying C-C bond forming reaction a question of debate.

---

[23] Tomoda, H.; Iwata, R.; Takahashi, Y.; Iwai, Y.; Ōiwa, R.; Ōmura, S.; *J. Antibiot.* **1986**, *39*, 1205 – 1210.

[24] Handa, M.; Ui, H.; Yamamoto, D.; Monma, S.; Iwai, Y.; Sunazuka, T.; Omura, S. *Heterocycles* **2003**, *59*, 497 – 500.

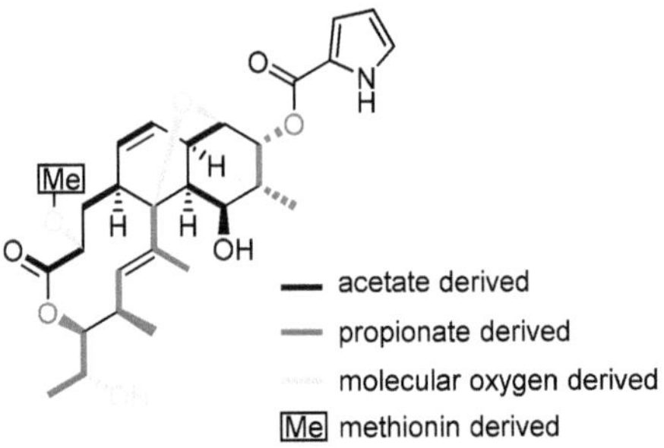

Figure 13: Nargenicin Biosynthesis

The polyketide nature of nargenicin[13] and nodusmycin[25] was proven by two independent ^{13}C labelling studies, published in two back-to-back papers. These studies showed that nargenicin and nodusmycin are both nonaketides, built from 5 acetate and 4 propionate units, and one methionin derived methyl unit. (Figure 13) All acetate and propionate units are connected in a head-to-tail fashion. Further studies on the late state oxidation executed in an $^{18}O_2$-atmosphere, showed that the oxygens not originating from acetate or propionate at C-2, C-8/13 and C-18 are indeed derived from molecular oxygen.[26] This information, put together, excluded both epoxide-olefin cylcisations and aldol reactions to be responsible for the formation of the *cis*-octalin carbacycle. Instead, an intramolecular Diels-Alder between a C-4 – C-7 *E,E*-diene and a C-12 – C-13 *E*-dienophile in the polyketide chain **24** was proposed, based on the *syn*-relationship of the protons at C-4 and C-7. The Diels-Alder product **25** is then further functionalised in late stage oxidation reactions.

---

[25] Snyder, W. C.; Rinehart, K. L. *J. Am. Chem. Soc.* **1984**, *106*, 787 – 789.
[26] Cane, D. E.; Yang, C.-C. *J. Antibiot.* **1985**, *38*, 423 – 426.

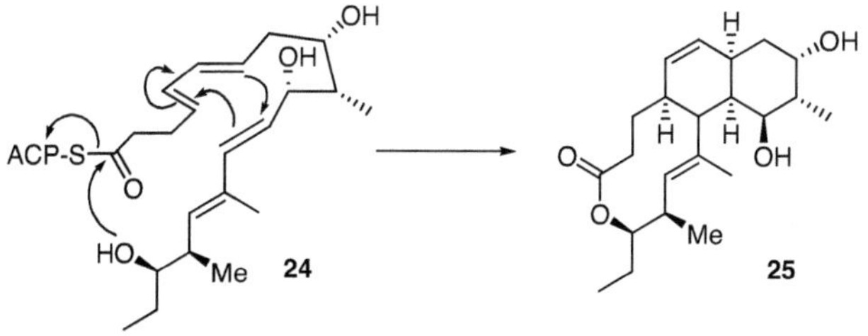

Scheme 1: Biosynthetic intramolecular Diels-Alder pathway

This mechanism could be proved by the incorporation of an appropriately ^{13}C-labelled fragment into nargenicin. It was shown that proposed polyketide chain intermediates, when administered as the *N*-acetylcysteamine thioester, could be incorporated into the nargenicin biosynthesis.[27] Therefore, ^{13}C-labelled diene **26**, when administered under carefully controlled fermentation conditions, could be incorporated into nargenicin.[28] (Scheme 2)

This shows diene **26** not only to be an active intermediate in the nargenicin biosynthesis, but also that the stereochemistry and oxidation state are set prior to the chain elongation. However, the actual oxidation state immediately before the Diels-Alder reaction remains unresolved – this question could be solved by deuterium labelling at C-11. However, without any disproof, an intramolecular Diels-Alder reaction remains the working hypothesis so far – be it using an unactivated dienophile as depicted in Scheme 1, or a related activated one. This in turn brings up the question if the the intramolecular Diels-Alder reaction is enzyme-catalysed, so to say, involves a "Diels-Alderase", or occurs spontaneously after macrolactonization under biological conditions.[29]

---

[27] Cane. D. E.; Ott, W. R. *J. Am. Chem. Soc.* **1988**, *110*, 4840 – 4841; Cane, D. E.; Tan, W. T.; Ott, W. R. *J. Am. Chem. Soc.* **1993**, *115*, 527 – 535.

[28] Cane, D. E.; Luo, G. *J. Am. Chem. Soc.* **1995**, *117*, 6633 – 6634.

[29] For reviews on biosynthetic Diels-Alder reactions, see: a) Williams, R. M.; Stocking, E. M. *Angew. Chem. Int. Ed.* **2003**, *42*, 3078 – 3115; b) Oikawa, H.; Tokiwano, T. *Nat. Prod. Rep.* **2004**, *21*, 321 – 352.

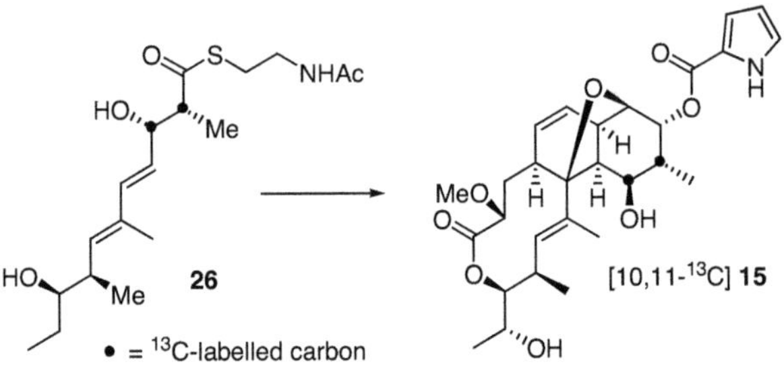

Scheme 2: Incorporation of a ^{13}C-labelled chain elongation intermediate

Biosynthetic studies executed on coloradocin[30] showed the "norther" half of this molecule to be derived from a polyketide synthase pathway as well. The "southern" half however, did not give an equally consistent picture. Two acetates appear to be connected tail-to-tail, which is attributed to the presence of an active tricarboxylic acid cycle, in which succinate is formed from two acetate units. Although the formation of the C-29 to C-32 carbon from succinate seems possible, feeding experiments with C-2, C-3 ^{13}C labelled succinate show that only C-26 to C-28 are of succinate orgin, whereas C-29 to C-32 are synthesised in an uncommon pathway from acetate. No ^{18}O-labelling studies have been conducted; hence the origin of the oxo-bridge oxygen remains unclear. Such studies could have shed some light on the origin of this 'abnormal' C9 - C13 oxo bridge.

---

[30] McAlpine, J. B.; Mitscher, L. A.; Jackson, M.; Rasmussen, R. R.; van der Velde, D.; Veliz, E.; *Tetrahedron* **1996**, *52*, 10327 – 10334.

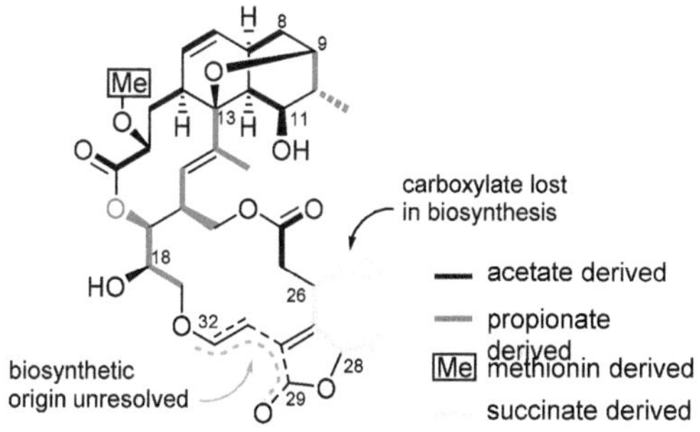

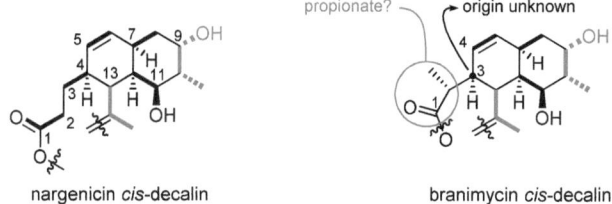

Figure 14: Coloradocin biosynthesis

No biosynthetic studies have been carried out on branimycin (**14**), however the fact that the macrolactone ring contains one carbon less brings up the question of its biosynthesis, as only a limited number of building blocks are normally incorporated into polyketides. (Figure 15)

Figure 15: Biosynthetic Differences between nargenicin and branimycin

In the case of nargenicin it was proven that starting with C-19 – C-17, the molecule is assembled stepwise. The similarity between branimycin and nargenicin from C-4 upwards suggests that a similar biosynthetic pathway is active in the synthesis of branimycin – therefore also the diene system should be constructed from acetate units. However, the C-1 to C-3 unit of branimycin does not fit in this picture, as this one-carbon shortened chain cannot be constructed from 2 acetates as in nargenicin. This might be explained as follows:

- A skeletal rearrangement takes places after the assembly of the full macrolactone chain – in this case, the branimycin C-2 methoxymethylen group would correspond to nargenicin C-2 or C-3.

- The branimycin C-1 to C-3 unit is introduced as one piece in a head-to-head coupling of methylmalonate to the polyketide chain.
- The branimycin C-1 to C-5 fragment is constructed from pyruvate (C-3 to C-5) and propionate (C-1 to C-2).

However, these explanations are purely speculative; furthermore, they are all based on the assumption that the branimycin C-2 methylen oxygen is introduced in a late stage of the biosynthesis, similar to the introduction of the C-2 methoxygroup in nargenicin.

2.4 Comparison and conclusion

Although 26 years have passed since the structure elucidation of nargenicin A_1 (**15**), some of its metabolites are still unknown, as their structure was not published in the patents that described their production. However, later biosynthetic studies showed that some minor metabolites were produced in the fermentation of *Nocaridia argentinensis* ATCC 31306, namely nodusmycin (**17**), 18-*O*-acetylnargenicin, 18-*O*-acetylnodusmycin, and isonargenicin. The formation of these co-metabolites, however, seemed to be strongly dependent on the fermentation conditions – two different research groups using different vegetative media observed different products.[13,25] For reasons of availability, the most biosynthetic studies were conducted on the stem *Nocaridia argentinensis* ATCC 31306 which produces nargenicin A_1.

The structures of the nargenicin family could be assigned by comparison with the NMR data available for nodusmycin, whose relative configuration could be determined undoubtedly by X-ray analysis. Its absolute stereochemistry was determined by the nonemperical CD exciton chirality method[26] and was later assigned in analogy.

The stereochemistry of coloradocin was assigned in a completely independent manner by extensive Mosher studies in combination with molecular mechanics studies, thus proving the stereochemistry for C-1 to C-18 previously proposed in analogy to the nargenicin family.

The relative stereochemistry of branimycin, on the other side, was assigned exclusively relying on NMR analysis. This assignment seems quite reliable for the rigid *cis*-octalin ring. However, as no NMR data is available for branimycin's closest relative in the nargenicin family, nargenicin B_1, a direct comparison of the configuration on the side-chain is quite difficult. Especially, as the macrolactone ring in branimycin is contracted to a 9-membered ring, direct comparison of the coupling constants is severely hampered, as the ring conformation of the 9-membered lactone ring might be considerably different to the nargenicin 10-membered macrolactone. Finally, it should be noted that the stereochemistry at C-2, C-14, C-15 and C-17 is derived from NOE interactions, which can be ambiguous on larger ring systems. Most notably, the configuration at C-17 (based on the assumption of hydrogen bonding of the 17-hydroxy function to the lactone carbonyl) is opposed to that of the nargenicin C-18, which makes it unique, as the configuration of this stereocenter had been consistent for all members of the nargenicin family so far.

However, as the NMR spectroscopy data is the only source of structural information, it has to be accepted as long as there is no contradiction from synthesis or related natural products isolations.

3 PREVIOUS SYNTHETIC WORK

3.1 Approaches towards the nargenicin family

3.1.1 Kallmerten's synthesis of 18-desoxynargenicin A_1

The first synthetic efforts towards the total synthesis of nargenicin A_1 were reported in 1984 by Kallmerten.[31] Retrosynthetically, the authors split the molecule into two parts, the "northern" *cis*-octalin unit and the polyketide side chain. The octalin-skeleton was synthesised in a racemic manner by intermolecular Diels-Alder reaction of 1-(trimethylsilyloxy)-butadiene and *p*-benzoquinone to yield the *endo* adduct **30**. (Scheme 3) DIBAL-H reduction and subsequent protection of the 1,3-diol gave the acetonide **31**, which could be converted to **32** by an S_N2' displacement with a methyl cuprate followed by acidic hydrolysis. The allylic alcohol could be oxidised selectively, and the remaining free hydroxy-group was protected as a MOM-ether. Regio- and stereoselective epoxidation of the non-conjugated double bond was achieved with *m*CPBA. Epoxide **33**, when treated with methyl magnesium bromide as a test reaction, furnished first a tertiary alcohol from the ketone, which then cyclised with the epoxide to yield the oxo-bridged bicycle **34**.

---

[31] Kallmerten, J. *Tetrahedron Lett.* **1984**, *25*, 2843 – 2846.

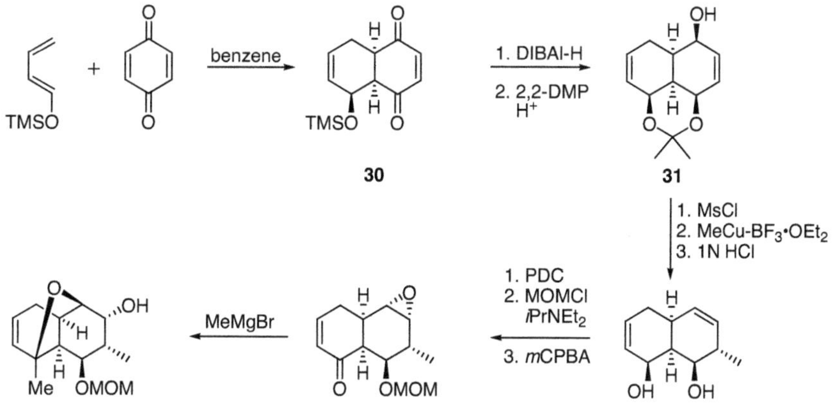

Scheme 3: Kallmerten's synthesis of the "northern" fragment **33**

The total synthesis of 18-desoxynargenicin was accomplished by the same authors 4 years later.[32] The addition of the non-racemic side chain **38**, after metallation with *t*-BuLi, (Scheme 4; previously published by Corey in the synthesis of erythronolide B[33]), led to a mixture of two diastereomers, as *cis*-octalin **33** was employed in racemic form. Both diastereomers had to be converted further, until structural assignment by X-ray analysis was possible on a crystalline intermediate. (Scheme 5) After identification of the correct MOM-protected diastereomer **41**, allylic oxidation yielded the enone in poor yield (39 %), to which the C1 to C-3 chain was attached by 1,4-addition. *In situ* trapping of the enolate gave the vinyl-phosphoimidate, which could be deoxygenated to fully functionalised octalin **42** under Birch conditions, although again in only poor yields. Cleavage of the C-1 acetal protecting group under mild conditions, followed by Jones-oxidation and methylation, led to C-1 methyl ester. α-Oxidation at C-2 could be

---

[32] Plata, D. J.; Kallmerten, J. *J. Am. Chem. Soc.* **1988**, *110*, 4041 – 4042.

[33] Corey, E. J.; Trybulski, E. J.; Melvin, L. S.; Nicolaou, K. C.; Secrist, J. A.; Lett, R.; Sheldrake, P. W.; Falck, J. R.; Brunelle, D. J.; Haslanger, M. F.; Kim, S.; Yoo, S. *J. Am. Chem. Soc.* **1978**, *100*, 4618 – 4620.

achieved using Vedejs' reagent;[34] after methylation of the C-2 oxygen, the diastereomers **43a** and **43b** could be separated. Identification of the right diastereomer **43a** was achieved by comparison with the degradation product of 18-deoxynargenicin by hydrolysis, esterification and MOM-protection of the secondary alcohol.

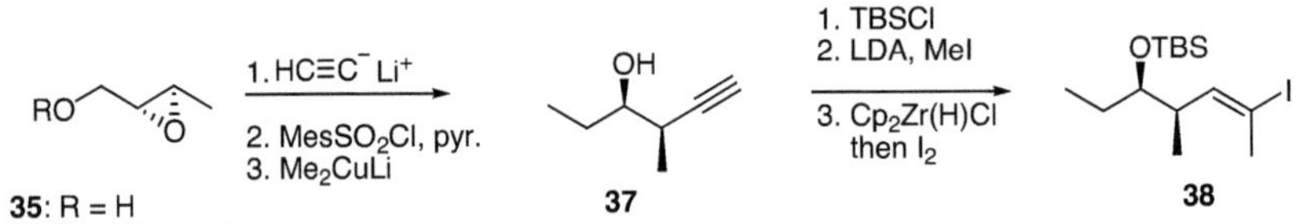

35: R = H
36: R = CMe2OMe

Scheme 4: Synthesis of polyketide side-chain **38**

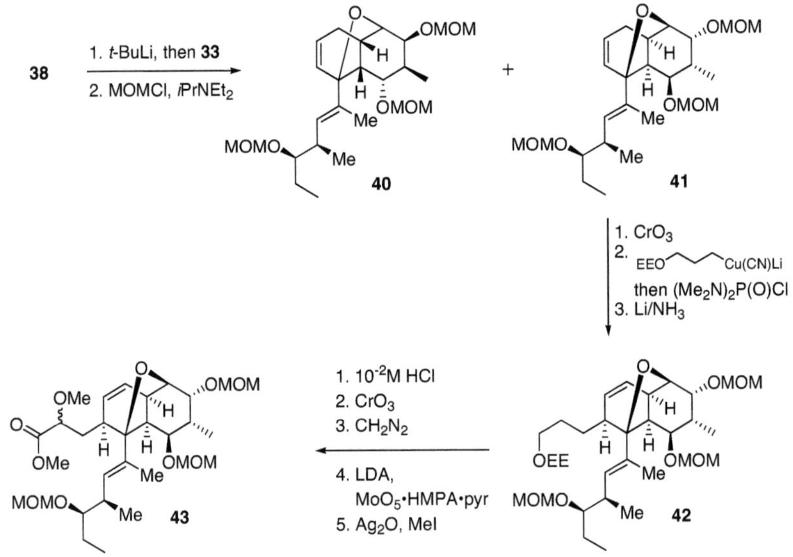

Scheme 5: Diastereomeric resolution and addition of the C1-C3 chain

[34] a) Vedejs, E. *J. Am. Chem. Soc.* **1974**, *96*, 5966; b) Vedejs, E.; Telschow, J. E. *J. Org. Chem.* **1976**, *41*, 740 – 741.

With the right diastereomer **43a**, the synthesis could be completed by global deprotection and final macrolactonization using the thiopyridyl procedure developed by Corey.[35] (Scheme 6) It was noted that only one C-2 epimer (the desired one) underwent cyclisation. Final formation of the 2-pyrrole-ester led to 18-desoxynargenicin A_1 (**16**), thus completing the up to date only total synthesis of a member of the nargenicin family.

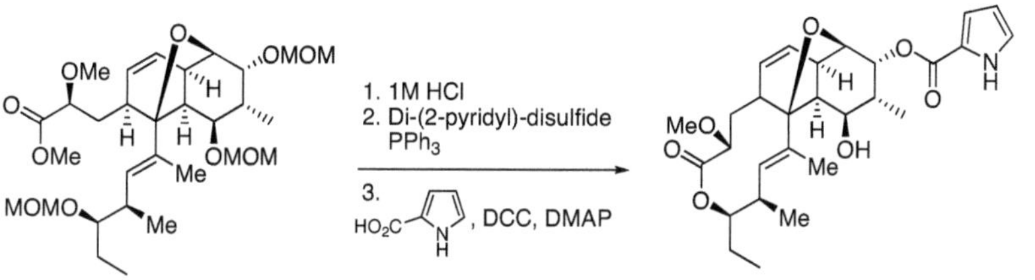

Scheme 6: Macroactonization and esterification

3.1.2 Kallmerten's approach towards 18-oxygenated nargenicins

Based on their work on the translation of chiral information in acyclic systems,[36] the Kallmerten group sought to prepare to C-16 to C-18 stereocenters exclusively under substrate control from already installed functionality.[37] To prepare a handle for the introduction of this functionality, alkylation of **33** was achieved with 1-lithio-1-ethoxy-ethene, which, after MOM-protection of the secondary alcohol, was hydrolysed to methyl ketone **44**. (Scheme 7) Condensation with acetaldehyde led to crotyl derivative **45**. After chelate-controlled 1,2–addition of MeLi to the enone, the resulting tertiary allylic alcohol could be alkylated with (chloromethyl)-oxazolidine **46** to give ether **47**. Treatment with LDA engaged a fully diastereoselective [2,3]-Wittig rearrangement, which set the stereochemistry at C-16 and C-17. Cleavage of the oxazolidine was accomplished by *N*-methylation and subsequent basic hydrolysis. The resulting free acid

---

[35] Corey, E. J.; Nicolaou, K. C. *J. Am. Chem. Soc.* **1974**, *96*, 5614 – 5616.
[36] Wittman, M. D.; Kallmerten, J. *J. Org. Chem.* **1988**, *53*, 4631 – 4633.
[37] Rossano, L. T.; Plata, D. J.; Kallmerten, J. *J. Org. Chem.* **1988**, *53*, 5189 – 5191.

was methylated and the C-17 hydroxyl protected as the MOM-ether **48**. To make this synthesis a formal synthesis of 18-desoxynargenicin A_1, the C-17 carboxylic ester was reduced to the alcohol, tosylated and then alkylated with dimethyl cuprate in a Schlosser-Fouquet coupling to give key intermediate **41**.

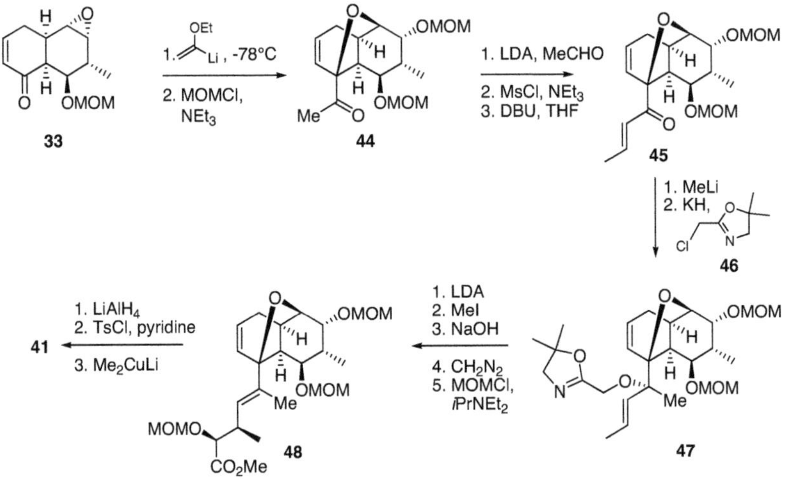

Scheme 7: Side-chain installation *via* [2,3]-Wittig rearrangement

Although this reaction sequence provided a rational and diastereoselective approach to selectively introduce all stereogenic centers (except that at C-2), it was longer than the previous one, and lacked the possibility of a chiral resolution, as the synthesis was still racemic. On the other hand, this approach allowed for the first time the installation of the oxygen-functionality in the 18-position present in Nargenicin A_1 (**15**). This was accomplished by methylation of the free acid intermediate **49** with MeLi, followed by chelation-controlled reduction to the *anti*-alcohol, which in turn was silylated to give orthogonally protected **50**.[38] (Scheme 8)

---

[38] Rossano, L. T. *Ph. D Thesis*, Syracuse University, **1990**.

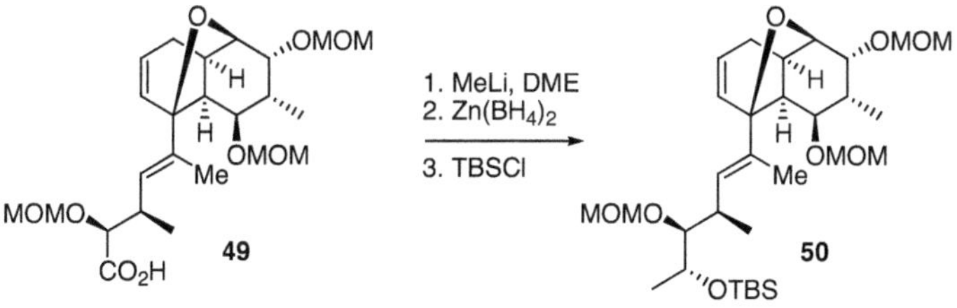

Scheme 8: Installation of the C-18-hydroxy function

Conversion of intermediate compound **50** to nargenicin A_1 would take about 11 steps (*cf*. Schemes 5,6). These requisite steps, however, were not performed.

3.1.3 Jones' approach to the oxo-bridged *cis*-octalin

Jones *et al*. published a model study on the construction of the oxo-bridged *cis*-octalin present in the nargenicin family.[39] In contrast to Kallmerten, they chose to construct this bicycle in an intramolecular Diels-Alder (IMDA) approach – a strategy that turned out to be biomimetic (*cf*. section 2.3). On a test system, they observed that trienone **52**, available from commercially available material in 2 steps, followed by C_2-elongation, underwent an *exo*-cycloaddition to give *cis*-octalin **53**. (Scheme 9) To furnish the desired oxo-bridge, the methine group α to the carbonyl had to be oxidised – under presumed inversion of configuration of the acetyl group. Unfortunately, the acid-catalysed dehydration of **54** to the oxo-bridged product concurred with a skeletal rearrangement, presumably due to the preferential formation of a hemiacetal, followed by a pinacol-type rearrangement/ring expansion. The authors reported no further efforts to suppress this rearrangement and form the desired oxo-bridge instead.

---

[39] Jones, R. F.; Tunnicliffe, J. H. *Tetrahedron Lett.* **1985**, *26*, 5845 – 5848.

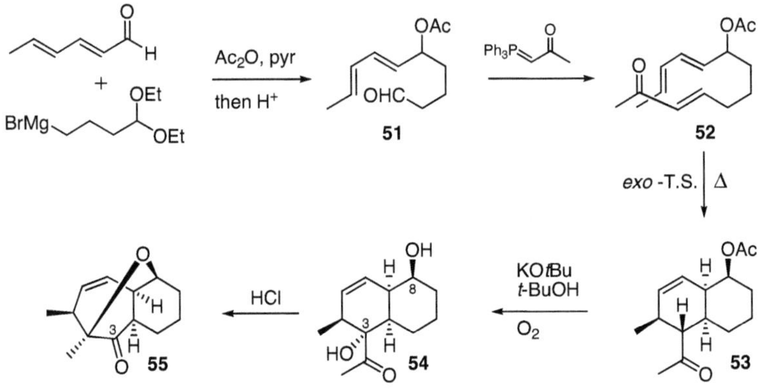

Scheme 9: Jones' access to the oxo-bridged octalin

3.1.4 Roush' IMDA approach to the *cis*-octalin core

Similar to the retrosynthetic considerations by Jones,[39] Roush *et al.* envisioned to construct the *cis*-octalin core of nargenicin in an IMDA reaction.[40] Based on previous work they proposed the reaction to proceed through *endo* transition state **57A**, which should give the desired stereochemistry present in nargenicin. (Figure 16)

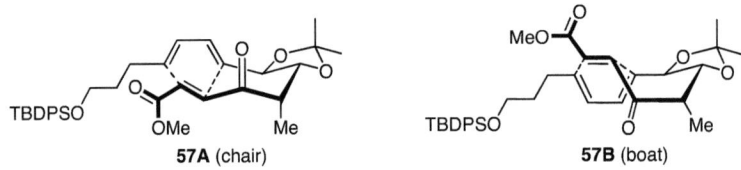

Figure 16: Possible *endo* IMDA transition states

---

[40] a) Roush, W. R.; Coe, J. W. *Tetrahedron Lett.* **1987**, *28*, 931-934; b) Coe, J. W.; Roush, W. R. *J. Org. Chem.* **1989**, *54*, 915 – 930.

Using their tartrate ester crotylboronate methodology,[41] the authors synthesised internally acetonide protected triol **58**. (Scheme 10) After ozonolysis, oxidative work-up and protection as the methyl ester **59**, the primary alcohol was oxidised to the aldehyde and then olefinated with Wittig-reagent **60**. Isomerisation with I_2 then delivered the *E,E*-diene **61**. The synthesis of the dienophile was accomplished by treatment of **61** with lithiated methyl dimethyl phosphonate, which was then condensed with methyl glycolate under modified Roush-Masamune conditions. Upon heating, **57** underwent a clean cylcoaddition to yield *cis*-octalin **62**.

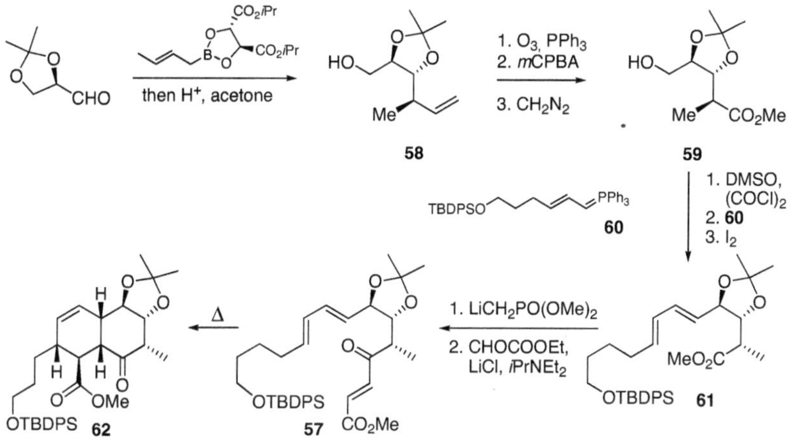

Scheme 10: Roush' IMDA approach

This outcome was most surprising, as this indicated the reaction to go through the boat-like transition state **57B**, opposed to the expected chair transition state **57A**. While the preference of 1,7,9-decatrien-3-ones to form *cis*-fused bicycles was known, the basis of this preference was not. To get a deeper insight into the origin of this preference, a nearly complete series of diastereomers of **57** prepared was prepared. As the authors had

---

[41] a) Roush, W. R.; Halterman, R. L. *J. Am. Chem. Soc.* **1986**, *108*, 294 – 296; b) Roush, W. R.; Hoong, J. L. K.; Palmer, M. A. J.; Park, J. C. *J. Org. Chem.* **1990**, *55*, 4109 – 4117; c) Roush, W. R.; Hoong, J. L. K.; Palmer, M. A. J.; Straub, J. A.; Palkowitz, A. D. *J. Org. Chem.* **1990**, *55*, 4117 – 4126.

assessed an IMDA reaction of a system bearing an acetonide linkage, this excluded a ring flip to the "other" chair conformation, which in turn allowed them to pinpoint the origin of stereocontrol in this reaction. Further extension to systems without the acetal tether showed that also in these reactions boat-like transition states play a significant role, although they now lead to *trans*-fused bicycles. Thus, the preferred conformation of such IMDA reactions involves a skew arrangement of the diene (which translates to the normal conformation of the $A^{1,3}$-strain model) to the chain combined with an eclipsed (*s-cis*) enone conformation, which results, in the absence of overriding steric factors, in the preference for the boat-like transition state **57B**.

3.1.5 Roush' biomimetic approach towards nargenicin A$_1$

After the mechanistically interesting, but synthetically non-productive IMDA approaches towards nargenicin, the Roush group turned their attention to access nargenicin in a transannular Diels-Alder (TADA) reaction.[42] This strategy represents a biomimetic approach, as the biosynthetic studies also propose a full elaboration of the polyketide chain, followed by macrolactonization and subsequent Diels-Alder reaction (see section 2.3 for details). This approach also promised to give some insight about a possible involvement of a 'Diels-Alderase' in the biosynthesis, as a spontaneous cycloaddition would put aside the necessity of a biocatalyst. However, in order to evaluate the general feasibility of such a transformation, the authors started their studies with a more fine-tuned substrate: A chiral *syn*-dioxolane was incorporated on the backbone, which had previously shown to give the highest preference for the *cis*-fused octalin under IMDA conditions.[40b] Furthermore, a bromine was introduced on the diene to increase the $A^{1,3}$-strain and in turn enhance the facial discrimination. For the simplest construction, the whole polyketide chain was assembled from 3 fragments of similar complexity.

---

[42] Roush, W. R.; Koyama, K.; Curtin, M. L.; Moriarty, K. J. *J. Am. Chem. Soc.* **1996**, *118*, 7502 – 7512.

The C-6 to C-12 building block **63** could be condensed with the C-13 to C-18 aldehyde **64**. The so produced enone **65** was then merged in a Suzuki-Miyaura coupling with the C-1 to C-5 boronic acid **66**. (Scheme 11)

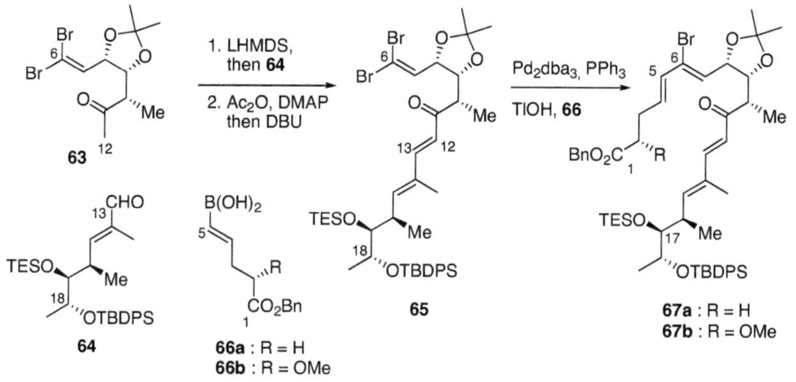

Scheme 11: Assembly of the C-1 to C-18 polyketide chain **67**

The polyketide chain **67a** was then deprotected to give the *seco*-acid which was then converted to macrolactone **68** under Yamaguchi conditions. (Scheme 12) Under thermal conditions, **68** converted selectively to *cis*-octalin **69** – however, the requisite oxo-bridge could not be introduced after the TADA reaction by allylic oxidation or remote functionalisation. Conducting the TADA reaction with a substrate bearing an oxygen substituent at C-13 did not give any product at all.

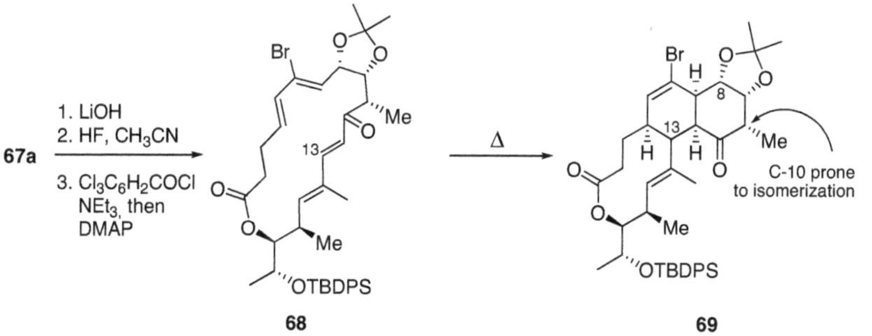

Scheme 12: TADA Reaction of 18-membered macrolactone **68**

However, no efforts have been undertaken to try this reaction with a truly biomimetic substrate – *i.e.* in protic media and a substrate bearing no protecting groups at C-9 and

C-18, C-11 reduced to the alcohol and no bromine at C-6. Even if such a Diels-Alder reaction is unlikely, it would give some insight if the Diels-Alder reaction can occur spontaneously in a non-enzymatic environment, or needs to be catalysed by a "Diels-Alderase".

3.1.6 Gössinger's approach towards nodusmycin

The synthetic approach to the *cis*-octalin core by Gössinger[43] is based on a similar disconnection as the total synthesis by Kallmerten; hence, the *cis*-decalin is assembled by an intermolecular Diels-Alder reaction between dimethoxy-tetrachlorocyclopentadiene (**70**) and *p*-benzoquinone (**71**) to assure the *cis*-relationship at the ring junction. (Scheme 13) In contrast to Kallmerten, the synthesis could be rendered chiral by desymmetrisation of the intermediate *meso*-compound to chiral tricycle **72**, which was then further converted by 1,4-addition followed by an α-oxidation to give **73**. In a radical dehalogenation reaction, a transannular lactol bridge was formed – changing to a benzyl protecting group then gave **74**. Addition of AlMe$_3$ followed by Lewis-acid catalysed fragmentation of the methylene bridge led to a 1,4-dicarbonyl, which could then be oxidised with FeCl$_3$ to give the *cis*-octalin **75**. The C-2 to C-4 side chain was then introduced by an intramolecular radical addition of an iodo-acetal to yield lactol **76**. Conversion of the methyl ketone to an exocyclic double bond, followed by opening and regioselective protection of the lactol gave **77**. The transannular oxygen bridge was introduced *via* an oxymercuration which produced the functionalised oxo-bridged core fragment **78**.

---

[43] a) Gössinger, E.; Graupe, M.; Zimmermann, K. *Monatsh. Chem.* **1993**, *124*, 965 – 979; b) Gössinger, E.; Graupe, M.; Kratky, C.; Zimmermann, K. *Tetrahedron* **1997**, *53*, 3083 – 3100.

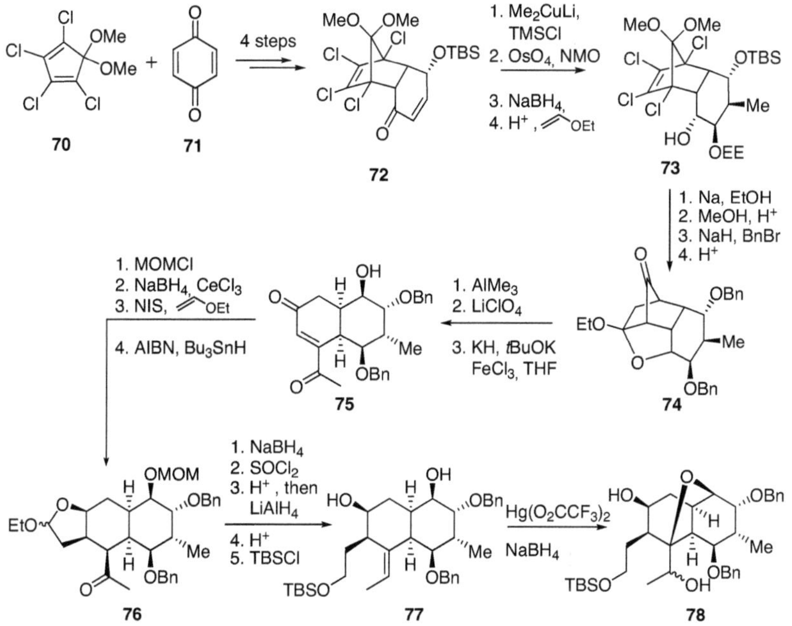

Scheme 13: Gössinger's approach towards nodusmycin

In further studies,[44] the authors also succeeded to attach the C-15 to C-19 side chain to diketone **75**. (Scheme 14) The C-14 methyl ketone of **75** could be converted to a vinyllithium species in a Shapiro process, which was then added to side chain fragment **79** (synthesised using Evans methodology). Barton deoxygenation now removed the C-15 hydroxy group under concomitant isomerisation of the double bond to give **80**.

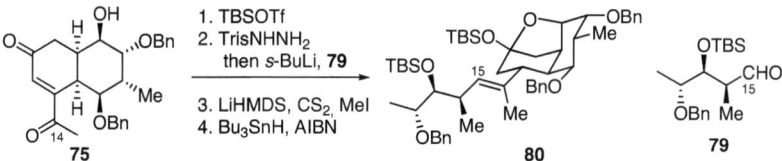

Scheme 14: Attachment of the C-15 - C-19 side chain **79** in Gössinger's synthesis

---

[44] Auer, E.; Gössinger, E.; Graupe, M. *Tetrahedron Lett.* **1997**, 6577 – 6580.

Despite considerable effort in the field and most of the important questions already solved, the Gössinger group has so far not finalised the total synthesis of nodusmycin.

3.2 Approaches towards Coloradocin

Probably due to the high similarity to the nargenicins, there have been no independent approaches towards coloradocin, but some that are based on previous approaches towards nargenicin. Kallmerten published some efforts to convert cis-octalin **33** (obtained from the previously published sequence[31]) to the coloradocin core. It turned out, that in contrast to the C-8 to C-13 bridge, the C-9 to C-13 bridge could not be closed by nucleophilic attack on C-9. Instead, the authors had to olefinate **33** and then reduce the epoxide regioselectively to alcohol **81**. Subsequent oxyselenation of the double bond resulted in the formation of the correct oxygen bridge in **82**, spanning from C-9 to C-13.

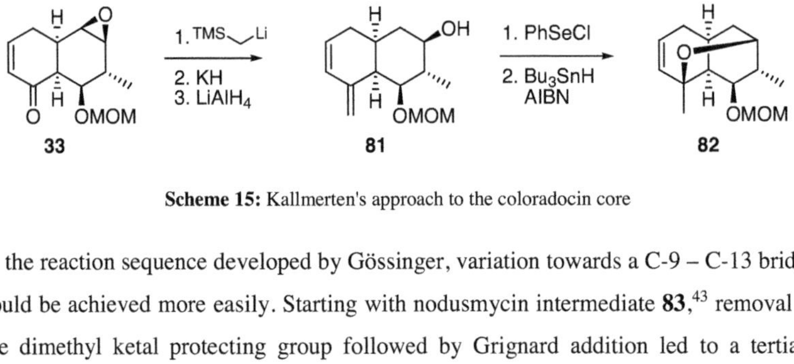

Scheme 15: Kallmerten's approach to the coloradocin core

In the reaction sequence developed by Gössinger, variation towards a C-9 – C-13 bridge could be achieved more easily. Starting with nodusmycin intermediate **83**,[43] removal of the dimethyl ketal protecting group followed by Grignard addition led to a tertiary alcohol, which after oxidation to the ketone, could be α-deoxygenated using SmI$_2$ to yield diketone **84**. Diastereoselective reduction and Lewis-acid induced fragmentation yielded appropriately functionalised cis-decalin **85**. No further efforts to form the C-9 – C-13 oxo bridge have been disclosed so far.

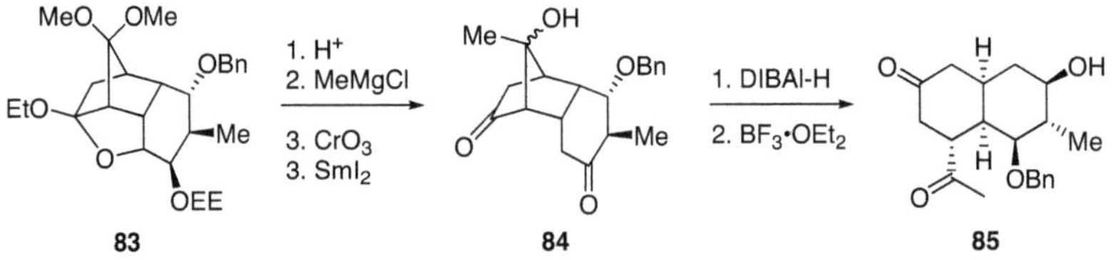

Scheme 16: Gössinger's approach to the coloradocin core

3.3 Other approaches towards branimycin in the Mulzer group

Three alternative routes towards the branimycin *cis*-octalin have been investigated previously in the Mulzer group.

3.3.1 Quinic acid based INOC approach

First, an approach starting from D-quinic acid (**86**) was investigated:[45] After some functional group interconversions to give protected cyclohexadienol **87**,[46] Mukaiyama aldol reaction followed by reduction and a Claisen-Ireland rearrangement furnished **88**. (Scheme 17) Conversion to the aldehyde and Z-selective Still-Gennari olefination gave α,β-unsaturated ester **89**. Formation of the oxime under basic conditions followed by oxidation to the nitrile oxide then led to a spontaneous intramolecular nitrile oxide cycloaddition (INOC) to give **90**. Unfortunately, a direct cleavage of the N-O bond was not possible due to a side reaction – therefore, the double bond had to be dihydroxylated first. Then, the cleavage of the isoxazolidine ring was possible, and addition of a nucleophile to the resulting ketone gave tertiary allylic alcohol **91**.

---

[45] a) Enev, V. S.; Drescher, M.; Kählig, H.; Mulzer, J. *Synlett*, **2005**, *14*, 2227 – 2229; b) Mulzer, J.; Castagnolo, D.; Felzmann, W.; Marchart, S.; Pilger, C.; Enev, V. S.; *Chem. Eur. J.* **2006**, *12*, 5992 – 6001.

[46] Murray, L. M.; O'Brien, P.; Taylor, R. J. K. *Org. Lett.* **2003**, *5*, 1943 – 1946.

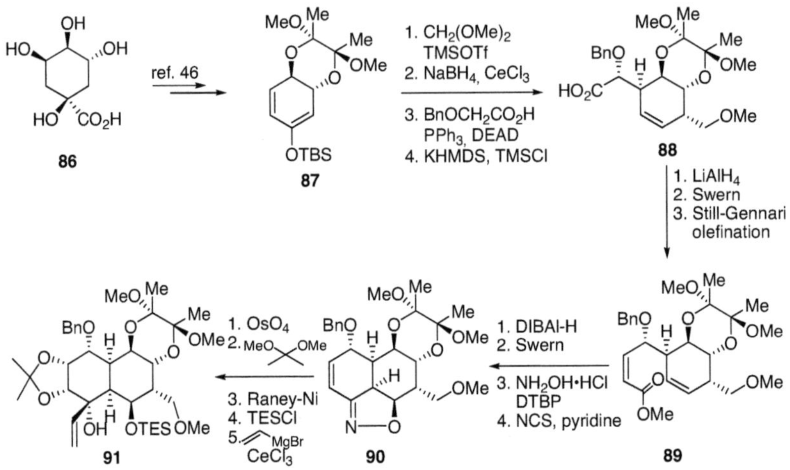

Scheme 17: Quinic acid based INOC approach

3.3.2 Lactone-templated INOC Approach

Independently from the 1st INOC approach, another [2+3]-cycloaddition route was evaluated.[45b] (Scheme 18) L-Ascorbic acid (**92**) derived chiral lactone **93** was used as central chiral building block. Copper-catalysed 1,4-addition and internal quench with acrolein delivered the C-10 alcohol as a 3:1 epimeric mixture, which was protected as TBS ether **94**. After liberation of the C-8 hydroxy function, it was converted to oxime **95**. Upon oxidation to the nitril oxide, the INOC reaction occurred only with the minor epimer of the C-10 TBS-ether to give **96**. Opening of the isoxazolidine, followed by reduction, gave the B-ring fragment **97**. For a more efficient INOC reaction, considerable effort was undertaken to exclusively access the *anti*-aldol product. As all attempts failed, this approach had to be abandoned.

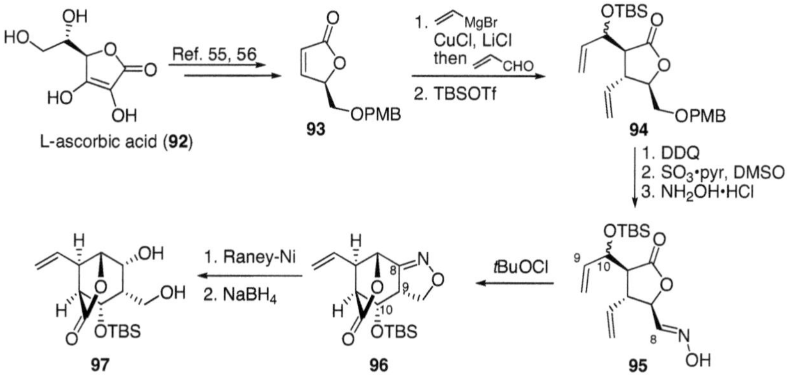

Scheme 18: Ascorbic-acid derived INOC approach by C. Pilger

3.3.3 Double-Claisen RCM Approach

Additionally, a Claisen-rearrangement-RCM approach was devised, using cyclohexenol **98**, available in analogy (*cf.* Scheme 17) from D-quinic acid derived starting material **88**.[47] (Scheme 19) On **98**, the O-PMB-glyoxylic ester could be introduced selectively, an Ireland-Claisen-rearrangement then yielded α-chiral acid **99**, which was reduced to the aldehyde and olefinated to give the terminal olefin **100**. Cleavage of the TES-ether, this stereocenter was inverted with benzyl-glyoxylic acid as nucleophile and another Ireland-Claisen rearrangement gave α-benzyloxy acid **101**. Again the carboxyl group was esterified, reduced and olefinated to the terminal vinyl group. This differentially protected double allylic alcohol was then subjected to a ring-closing metathesis (RCM) which yielded the *cis*-hexalin **102**. It was noted that only one epimer of the benzyloxy ether converted to the *cis*-hexalin under the given reaction conditions. After removal of the PMB-group and oxidation to the α,β-unsaturated ketone, epoxidation of the isolated double bond with *m*CPBA occurred from the convex face to yield selectively desired **103**.

---

[47] Marchart, S.; Mulzer, J.; Enev, V. S. *Org. Lett.* **2007**, *9*, 813 – 816.

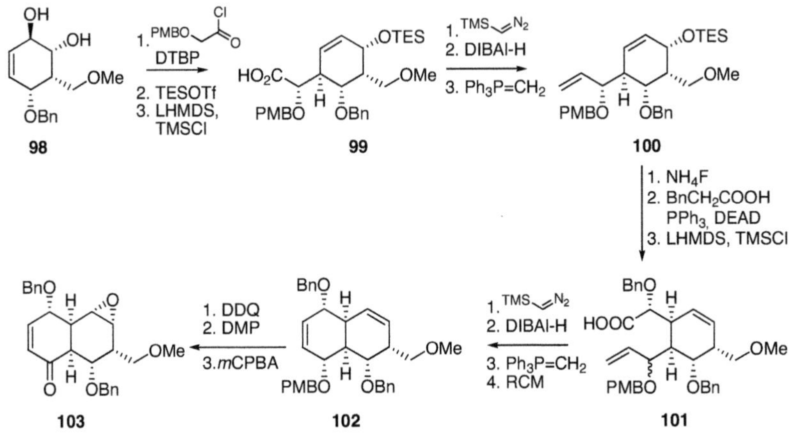

Scheme 19: Claisen-RCM approach towards the *cis*-octalin

During the course of the work described in this Ph.D thesis, this approach was still active and represents a point for comparison.

4 OUTLINE OF THE SYNTHETIC PLAN

The aim of this Ph.D thesis was to find a short and diastereoselective entry to the highly substituted *cis*-octalin core **105** of branimycin. (Scheme 20) The enantiopure synthesis should ideally transfer the chiral information from starting material exclusively in substrate-controlled transformations. We planned a convergent approach towards branimycin (**14**) – the molecule can be disconnected into two main fragments, the *cis*-octalin core fragment **105** and the metallated side-chain fragment **106**. Ideally, **105** should already contain suitable functionality for easy introduction of a metallated side chain **106**, which could then form the transannular oxo-bridge at a late stage of the synthesis. The only operations then left are the attachment of the C-2 methoxymethylen group and global deprotection to give the natural product branimycin **14**.

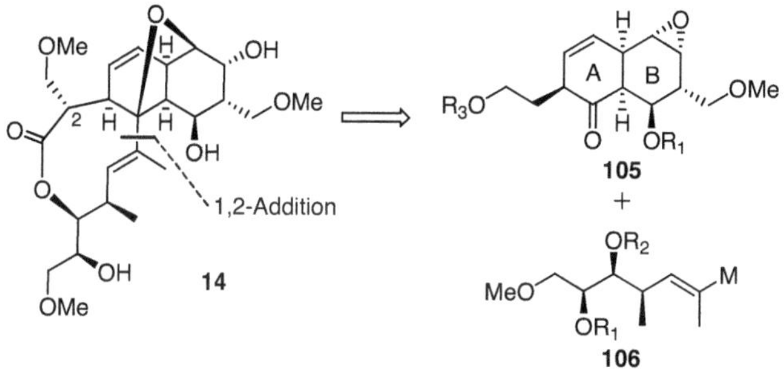

Scheme 20: Planned disconnection of branimycin (**14**) into *cis*-octalin (**105**) and side-chain (**106**)

Due to the considerable synthetic effort executed in the field of total synthesis of related natural products, we evalutated the results obtained by previous research in this field. In all approaches (but one) described in section 3, the *cis*-octalin core was constructed in a Diels-Alder reaction or a related cycloaddition.

In general, the approaches using <u>inter</u>molecular Diels-Alder reactions accessed the *cis*-octalin relatively quick (10 steps) when executed in a racemic manner – for asymmetric variations, considerably longer sequences were needed. A racemic synthesis seemed inacceptable given today's state-of-the-art in the synthetic field – however, to render such an intermolecular Diels-Alder reaction chiral without any additional desymmetrisation steps, an enantioselective Diels-Alder catalyst capable of transforming oxygenated dienes is necessary. However, there are only scarce examples in the literature of such reactions being carried out at all – a generally applicable catalyst with high asymmetric induction has not been put forward so far.

An intramolecular Diels-Alder reaction on the other hand, when applied to close the A-ring, produced *cis*-octalins with the ring-junction stereocenters set opposite to the expectation and precedence from the proposed biosynthesis. Stereochemical induction was easily possible in this case, as the chiral information on the chain connecting diene and dienophile translated efficiently into the newly formed stereocenters.

Evaluating the findings by our predecessors in this field, we thought to use the failed efforts by Roush[40] to our advantage, and access the nargenicin core in an IMDA reaction. In analogy to these findings, the desired product should be formed when the B ring is closed in a related IMDA reaction. A Diels-Alder product like **107** could then be converted to **105** by removal of the ester at C-10, epoxidation of the C-7 – C-8 double bond and conversion of the acetonide-diol to a double bond followed by an inversion of the C-5 alcohol and an Ireland-Claisen rearrangement to install the C-3 side-chain. (Scheme 21) A trienoate like **108** would be an ideal precursor for the IMDA reaction to give **107**.

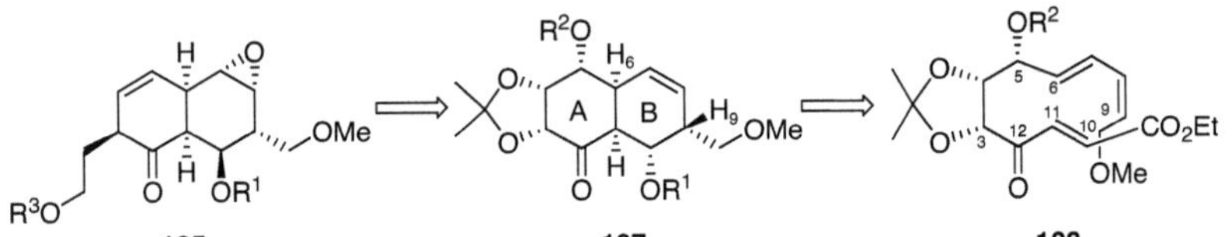

Scheme 21: Proposed IMDA reaction to access *cis*-octalin core

The efficient formation of IMDA product **107** is based on the rational design of the Diels-Alder precursor **108**. The diastereocontrol of this reaction should be exerted – inspired by the findings of Roush[40] – by an acetonide tether, which has shown to generate excellent diastereoselectivities on related systems. (Figure 17) On the B ring of **107**, H-6 and H-9 are in *anti* relationship – this in turn necessitates the diene in **108** to be either *E,Z* or *Z,E* configured. Molecular model inspections showed, that the steric requirements of the chain connecting diene and dienophile allow only an *E,Z* configured diene to undergo the desired IMDA reaction. The configuration of the C-5 hydroxyl should be beneficial for this reaction as well, partly because of the antiperiplanar alignment between the C-O bond and the newly formed C-6 – C-11 bond, and partly due to a conformational restriction of the diene's orientation based on the $A^{1,3}$-strain.

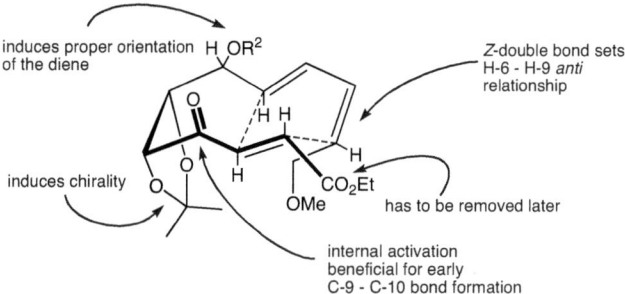

Figure 17: Rational design of the IMDA precursor **108**

The required chiral triene should be easily available from commercially available D-ribonolactone (**110**). (Scheme 22) After selective acetonide protection of the *cis*-diol, the primary alcohol should be oxidised and converted to the *E,Z*-diene **109** in two olefination steps. Opening of the lactone **109** with a lithiated phosphonate and subsequent C_2-elongation with ethyl glyoxalate should then give the IMDA-precursor **108**.

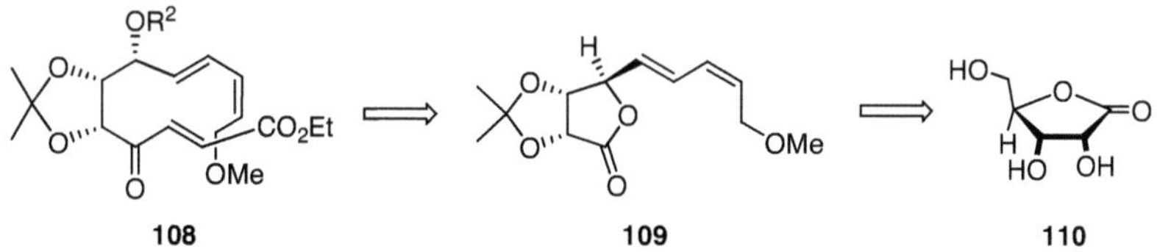

Scheme 22: Retrosynthesis of **108**

This sequence deemed fast and reliable enough to serve as an entry point into the branimycin *cis*-octalin. The absolute configuration is ascertained as the starting point is a chiral pool molecule – all subsequent stereocenters shall be constructed based on this chirality.

RESULTS AND DISCUSSION

Several approaches towards the highly substituted *cis*-octalin part of branimycin or related molecules have been described and evaluated in chapter 3; additionally, in chapter 4 a new retrosynthetic concept as a basis for this work has been proposed. Based on this strategy, 3 different approaches towards the core – fragment have been undertaken; each of them based on the knowledge about intrinsic molecular properties learned in the previous approaches. For a better understanding, all three approaches are summarised here:

1st Generation approach: This approach is based on the originally proposed retrosynthetic analysis. It is closely oriented on similar observations by Roush[40]– following these results, an acetonide tether is used for stereocontrol, assisted by a properly oriented C-5 OR^2 group. All these stereochemical features were installed for a most selective IMDA reaction – however, substantial functional group interconversions were required afterwards.

2nd Generation approach: In order to make the synthesis more convergent, the C-4 – C-5 double bond should already be incorporated before the IMDA reaction, thus avoiding the inversion of the alcohol and subsequent Ireland-Claisen rearrangement. In this approach, the chiral information comes from a lactone tether, which bears the additional advantage to contain the C-1 – C-3 side chain in the configuration needed for the natural product. Due to the conformational lock of the *trans*-γ-lactone, in combination with additional unsaturation on the diene-dienophile linkage, a useful stereocontrol could be expected.

3rd Generation Approach: In the 2nd Generation approach it was found that the IMDA reaction was not as easily accomplished as originally anticipated. Especially, the stability of the C-8 – C-9 Z-double bond gave rise to concerns about the feasibility of this approach. Hence, the idea arose to stabilise this double bond by incorporation into a macrolactone. Due to the additional conformational restriction imposed by the highly unsaturated macrolactone ring, we hoped for even higher asymmetric induction by the chiral lactol ring in the transannular Diels-Alder (TADA) reaction.

Convergent Side-Chain synthesis: For a rapid entry towards the side-chain stereotriade and easy variation of protecting groups, a new diastereoselective propargylation approach was developed. A nucleophilic propargylation of a glyceraldehyde derivative with an axially chiral silyl allene, led to selective formation of the required *syn*, *syn* stereotriade with variation of all protection groups possible.

5   1ST GENERATION APPROACH

In our first approach, we sought to access the branimycin *cis*-octalin core *via* an IMDA reaction of a triene derived from D-ribonolactone (**110**). Protection of the *cis*-diol to give the acetonide **111** was best achieved using 2,2-dimethoxypropane and catalytic amounts of I_2.[48] (Scheme 23) If harsher conditions were applied (large excess of 2,2-dimethoxpropane and camphor sulfonic acid as acidic catalyst), opening of the lactone and formation of the bis-acetonide was observed.[49] The oxidation of the ribonolactone C-5 alcohol turned out to be more problematic than originally anticipated: Oxidation attempts *via* the Swern protocol failed as well as Dess-Martin periodinane (DMP) oxidations under standard conditions. Closer inspections of the literature revealed a detailed purification method for the synthesis of aldehyde **113** using DMP (**112**),[50] which indicated this aldehyde to be of a very labile nature. Indeed, when the reaction was not quenched by addition of sodium bicarbonate, but only the reagent precipitated by addition of diethyl ether followed by filtration and evaporation of the solvents, crude aldehyde **113** could be isolated after one additional trituation step in good yield (91 %).

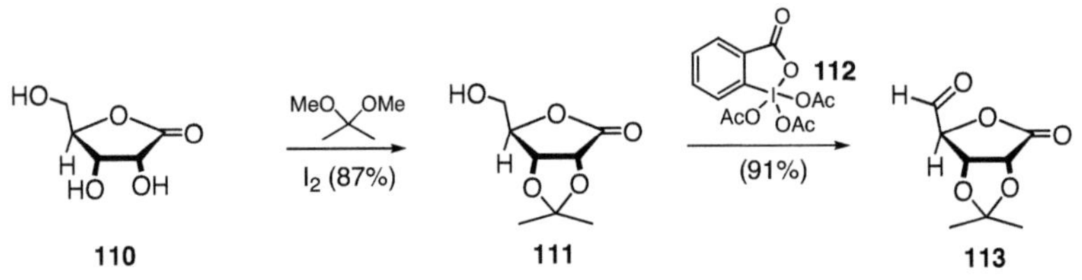

Scheme 23: Synthesis of the ribonolactone aldehyde **113**

We first planned to construct the diene moiety stepwise using the appropriate Wittig-reagents, which should deliver the desired double bond geometry. First, the *E*-double

---

[48] Raveendranath, P. C.; Blazis, V. J.; Agyei-Aye, K.; Hebbler, A. K.; Gentile, L. N.; Hawkins, E. S.; Johnson, S. C.; Baker, D. C. *Carbohydr. Res.* **1994**, *253*, 207 – 223.
[49] Barton, D. H. R.; Liu, W. *Tetrahedron*, **1997**, *53*, 12067 – 12088.
[50] Siozaki, M.; Arai, M.; Kobayashi, Y.; Kasuya, A.; Miyamoto, S. *J. Org. Chem.* **1994**, *59*, 4450 – 4460.

bond should be installed using a stabilised Wittig-reagent – in this case triphenylphosphoranylidene-carbaldehyde.

Upon reaction of **113** with triphenylphosphoranylidene-carbaldehyde at low temperatures in benzene, the desired α,β-unsaturated aldehyde **114** was produced only in low yields (25 %). (Scheme 24) Furthermore, all attempts to install the Z-double bond with Wittig-salt **115** under salt-free conditions, gave at best only ~10% yield of **116** as a mixture of isomers.

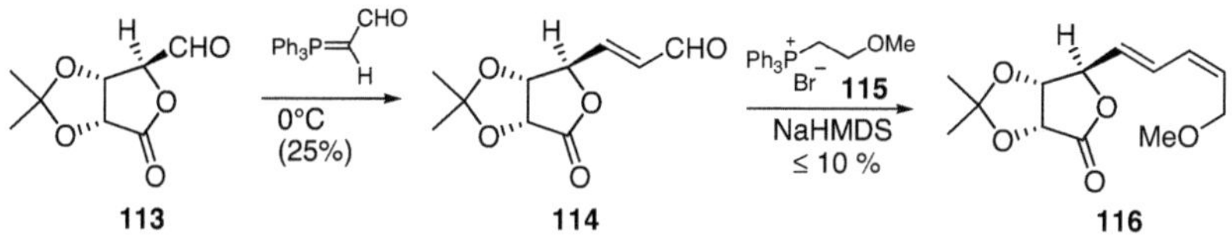

Scheme 24: Attempted E,Z-diene synthesis

An alternative approach to access the E,Z diene **116** involved a cross-coupling reaction to build the diene. However, when aldehyde **113** was treated with 3 eqivalents of iodoform and 6 equivalents of $CrCl_2$ in THF following the Takai-protocol,[51] the desired E-vinyl iodide **117** could only be obtained in trace amounts (~5 % yield), accompanied by some deprotected starting material; (Scheme 25) furthermore, it turned out that the appropriate Z-vinyl-boronates or -stannanes **118** required for the coupling are much harder to obtain than their corresponding E-isomers.

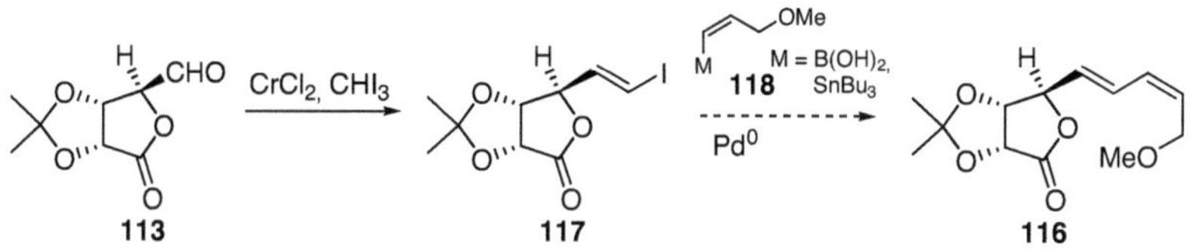

Scheme 25: Alternative E,Z-diene construction

---

[51] Takai, K.; Nitta, K.; Utimoto, K. *J. Am. Chem. Soc.* **1986**, *108*, 7408 – 7410.

At this point we realised that aldehyde **113** was apparently too unstable to be used efficiently for an olefination reaction. Therefore we thought to open the lactone beforehand and to construct **119** by introduction of the diene moiety in a nucleophilic attack on a one-carbon shortened aldehyde and avoid the unstable aldehyde **113**. Following the open-book effect as depicted in transition state **120**, the attack should occur primarily on the *si*-face, thus recreating the original stereochemistry. However, literature precedence indicated, that a direct C_1-degradation of **111** to a 1,4-dicarbonyl like **121**, would lead to another instable molecule, this time in terms of enantiomeric purity.[52] (Scheme 26) Therefore, a better combination would be the attack of metallated diene **122** on a chiral aldehyde **123** where the second carbonyl has been reduced beforehand. Aldehyde **123** should be accessible by opening of lactone **124** with an appropriate nucleophile **125** or **126**.

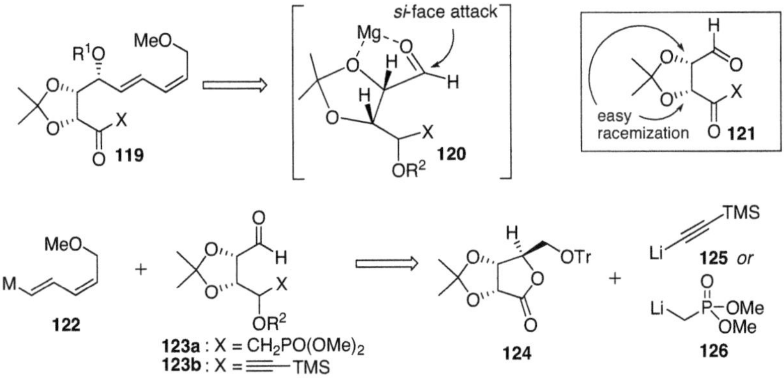

Scheme 26: Approach using a one-carbon shortened sugar moiety

We first investigated feasible approaches towards the side-chain **122**. Starting from propargylic alcohol, the alcohol was first methylated, followed by deprotonation of the alkyne with *n*-BuLi and quenching with iodine to give iodo-alkyne **127**. The conversion of the iodo-alkyne to the sensitive Z-vinyl iodide **128** was possible by reduction with diimide. However, to avoid overreduction to the iodoalkane, low temperatures and slow

---

[52] Rodriguez, J. B. *Tetrahedron* **1999**, *55*, 2157 – 2170.

generation of equimolar amounts of diimide had to be employed. Such conditions can be realised by addition of an equimolar amount of acetic acid *via* syringe pump to a buffered suspension of dipotassium-azodicarboxylate at 0 °C.[53] Although the Z-vinyl iodide **128** could be obtained in excellent yields (93%) in the first attempts, later efforts to scale this reaction up were difficult – partly because of experimental difficulty, and more important, because the quality of the dipotassium-azodicarboxylate could not be determined prior to its use.

Sonogashira-coupling of **128** with TMS-acetylene under standard conditions[54] produced the desired enyne **129** in nearly quantitative yield. However, subsequent removal of the acetylene TMS-group using TBAF revealed a flaw of the synthetic concept, as the product **130** was too volatile to allow useful separation from the solvent (THF). Therefore, the final hydrostannylation to yield the *E,Z*-diene **122a** was not be attempted.

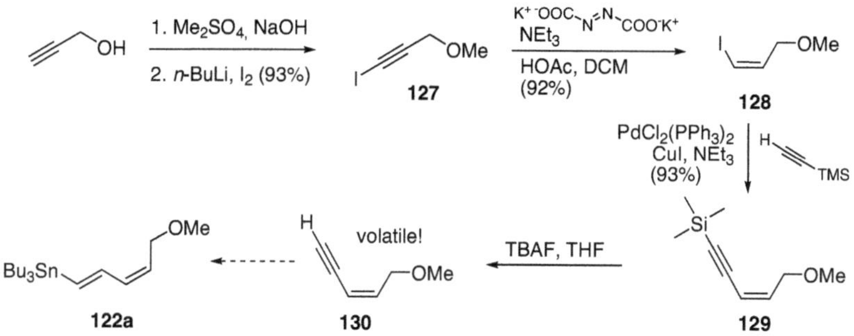

Scheme 27: Attempted construction of the *E,Z*-dienyl nucleophile **130**

To evaluate the overall strategy despite these petty hindrances, studies about the availability of the one-carbon shortened sugar fragment **123** were undertaken.

After protection of the primary alcohol **111** as the trityl ether **124**, first the ring-opening of the lactone with lithiated dimethyl-methylphosphonate was investigated. (Scheme 28) This reaction is reported to be cleanest when a second equivalent of LDA and 2

---

[53] a) For a review on diimide reductions, see: Pasto, D. J.; Taylor, R. T. *Org. React.* **1991**, *40*, 91 – 155; b) Hoffmann, R. W.; Hense, A. *Liebigs Ann.* **1996**, 1283 – 1288.

[54] Sonogashira, K.; Tohda, Y.; Hagihara, N. *Tetrahedron Lett.* **1975**, *16*, 4467 – 4470.

equivalents of TESCl are added to the reaction mixture, in order to trap both the secondary alcohol and the β-ketophosphonate enolate as TES-ethers.[55] However, with our system, no conversion to **131** could be observed under these reaction conditions.

As an alternative, we decided to introduce the required dienophile in the form of TMS-acetylene, which could later be reduced to the corresponding alkene. Such a dienophile would fit well into the overall synthetic scheme, as no decarboxylation would be necessary after the Diels-Alder reaction. The addition of 2 equivalents of lithiated TMS-acetylene worked very well at -78 °C and gave rise to an epimeric mixture of lactols **132** (74 %), accompanied by a small amount of de-silylated product (21 %).

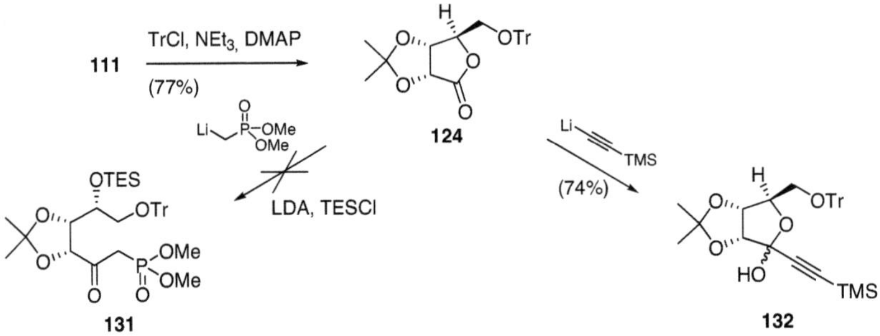

Scheme 28: Efforts towards the lactone ring-opening

All attempts to protect the ynone in its tautomeric form as the lactol acetate failed. (Scheme 29) Protection after work-up using an excess of reagent led to exclusive formation of vinyl acetate **133**, whereas milder *in situ* trapping may have led to formation of the desired lactol acetate **135**, but upon column chromatography this product rearranged to cyclic ketal **134** *via* proposed oxonium ion **136**.

---

[55] Ditrich, K.; Hoffmann, R. W. *Tetrahedron Lett.* **1985**, *26*, 6325 – 6328.

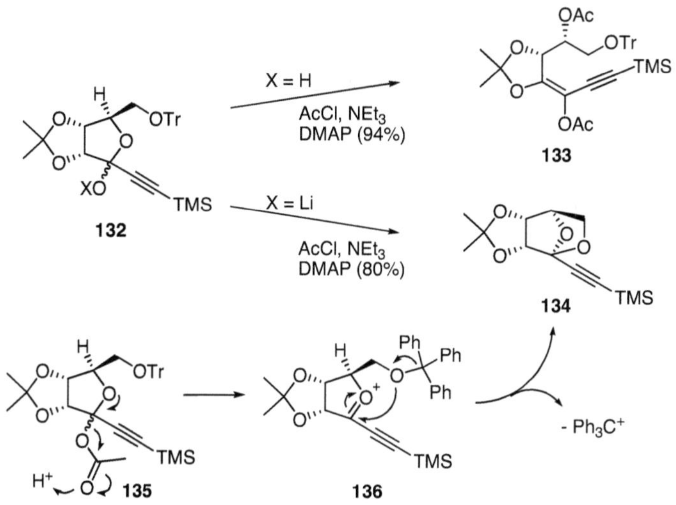

Scheme 29: Observed side-reactions upon acetate trapping

As the carbonyl could not be trapped as the lactol acetate, we decided to reduce it to the propargylic alcohol, hoping a kinetic differentiation would be possible afterwards. (Scheme 30) Reduction of the lactol **132** to the propargylic alcohol **137** was possible with NaBH₄, albeit a significant amount of desilylated side-product was formed. Buffering the reaction with NH₄Cl and limiting the reaction time led to nearly exlusive formation of **137**. Unfortunately, the desired selective TES-protection of the diol **137** leading to **138** did not work as good as anticipated – a nearly 1:1 mixture of regioisomers was produced.

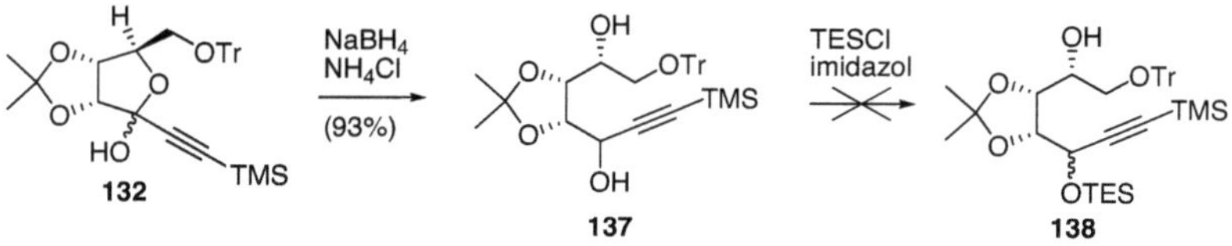

Scheme 30: Attempted final differentiation of diol **137**

At this point we had to reconsider our strategy: As the olefination of the ribonolactone C-5 aldehyde had turned out to be cumbersome – furthermore, the alternative nucleophilic attachment of the diene was hampered by difficulties in the synthesis of the diene nucleophile. As finally also the conversion of the lactone to the terminal enone was not fruitful, we decided to turn our attention to a new approach using another chiral pool starting material.

6   2ND GENERATION APPROACH

6.1   Concept

Instead of using D-ribonolactone as a chiral scaffold and following then the *cis*-octalin synthesis by Roush, we decided to investigate a more original strategy. This would give our synthesis more independence and a higher momentum due to a faster assembly of the core system: Before, a chiral acetonide tether spanned the positions 3 and 4 to induce the stereochemical information into the IMDA reaction – however, this concept necessitated that the C-4 – C-5 double bond had to be introduced later in the synthesis. On the other hand, if the chiral tether spanned C-3 – C-12 instead, an additional double bond could be present prior to the IMDA reaction. (Figure 18) As now two endocyclic double bonds are present in the molecule, the C-7 – C-8 double bond needs to be addressed selectively. Such a differentiation should be possible by epoxidation directed by the C-9 hydroxymethylen group.

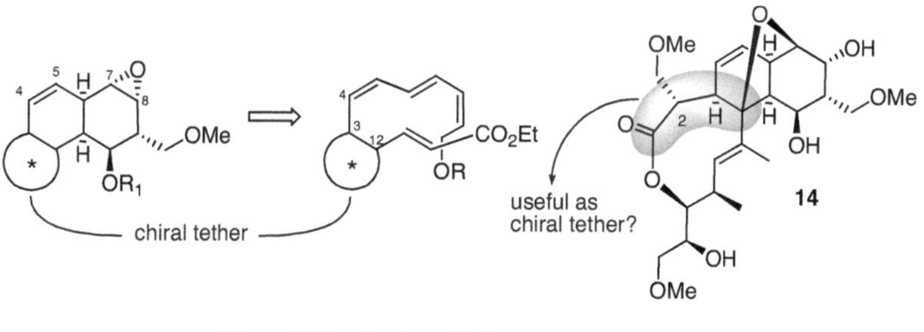

Figure 18: Implications of a C-3 – C-12 chiral tether

Looking at the target molecule branimycin (**14**) itself, such a tether can be seen in the form of the C-1 – C-3 to C-12 chain, connected as a 5-membered lactone ring, bearing the carbonyl functionality at C-1 (as present in the natural product), and the hydroxy-function at C-12 (which has to be oxidised to the ketone prior to attachment of the side-chain). Furthermore, the methoxy-methylene group at C-2 could already be installed at this early stage.

Ideally, the stereochemistry at C-3 in the Diels-Alder precursor **140** should be the one of the final product, whereas variation of the C-12 hydroxy stereocenter is possible, as this stereocenter will be destroyed later. (Scheme 31) Inspections of a molecular model revealed that a lactone carrying the substituents in a *cis*-relationship would be too flexible to be of use as a chiral tether – a *trans*-relationship on the other hand separates both diene and dienophile more efficiently and suggests to give preference for the desired transition state **140A**. To allow the reaction to go through the electronically favoured *endo*-transition state, the dienophile in **140** should bear a Z-double bond. (Scheme 31)

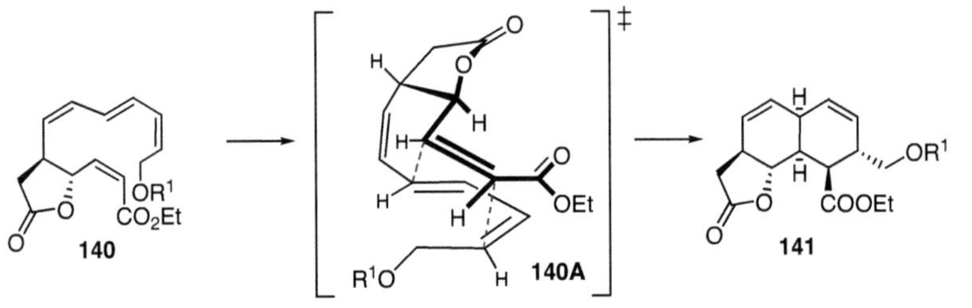

Scheme 31: Proposed IMDA reaction of lactone tethered tetraene **142**

This approach, compared to the previous one, would avoid most of the delicate steps after the pivotal IMDA reaction: Using the lactol-tethered approach, the C-3 – C-4 double bond generation, the inversion of the C-5 alcohol and the Ireland-Claisen rearrangement to install the C-4 – C-5 double bond and the C-3 side chain with the correct stereochemistry are obsolete – this corresponds at least to 5 steps saved. (Figure 19) Now, the only substantial transformation required after the IMDA reaction, besides the hydroxy-directed epoxidation of the C-7 – C-8 double bond, is the degradation of the ester group at C-10 to the alcohol. This transformation, however, was also necessary in the previous approach.

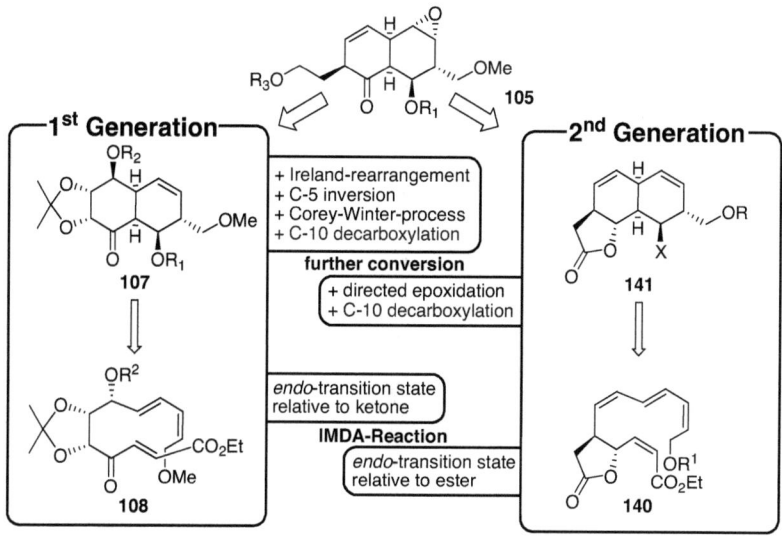

Figure 19: Comparison 1ˢᵗ Generation vs. 2ⁿᵈ Generation synthesis

6.2 Chiral tether synthesis

Inspired by previous studies by C. Pilger,[45b] we thought that the chiral tether building block would be most easily accessible starting from L-ascorbic acid (**92**). Following a literature procedure,[56] catalytic hydrogenation in water at 50 °C led to L-gulono-1,4-lactone (**144**) in good yields. (Scheme 32) Kinetic protection of the exocyclic diol led to monoacetonide **145**, which could then be cleaved with sodium periodate under carefully controlled conditions using a pH-electrode to yield *S*-glyceraldehyde acetonide **146**.[57] *In-situ* olefination with the stabilised Wittig-reagent led to a 75:25 mixture of the *Z*- (**147**) and *E*- (**148**) unsaturated esters.

---

[56] Andrews, G. C.; Crawford, T. C.; Bacon, B. E. *J. Org. Chem.* **1981**, *46*, 2976 – 2977.
[57] Hubschwerlen, C. *Synthesis* **1986**, 962 – 964.

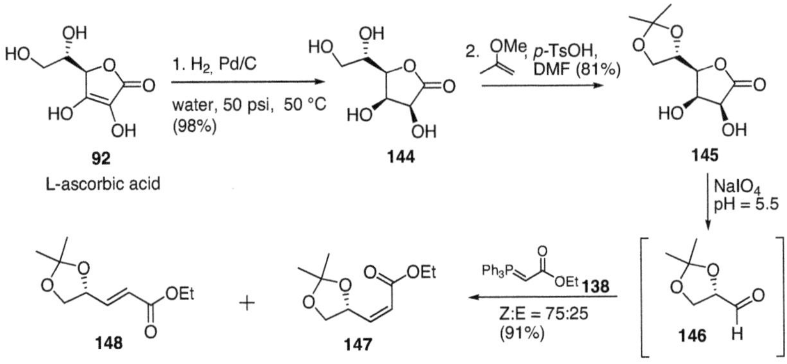

Scheme 32: Synthesis of chiral acrylates **147** and **148**

This surprising Z-preference can be explained by looking at the kinetics of such Wittig-reactions. (Scheme 33) Normally, when a stabilised Wittig-reagent is used, the *cis*-substituted oxophosphetane **149** formed initially (driven by the repulsion of phenyl group and R^1 – see **149A**) can undergo *retro*-cycloadditions to isomerise to the more stable *trans*-substituted oxophosphetane ring **150**, which upon expulsion of Ph₃PO then leads to the *E*-configured isomer. However, in the case of acetonide-glyceraldehyde derivatives it seems that either the initial *cis*-fused oxophosphetane ring is more unstable due to electronic effects of the dioxolane ring, or the steric repulsion between the dioxolane ring and the ester group dominates. In either case, formation of larger-than-normal amounts of the Z-acrylate **147** is observed.

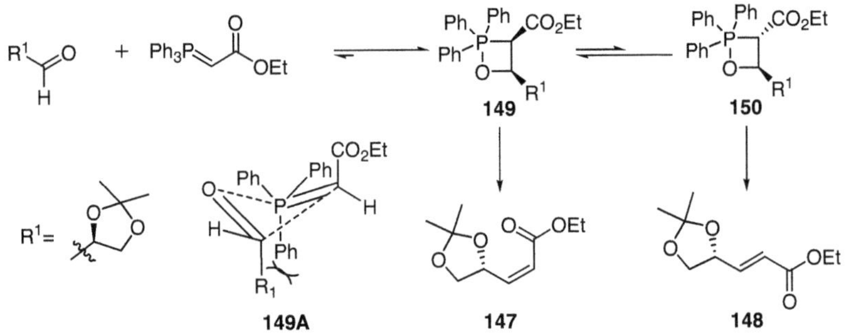

Scheme 33: Kinetic preference of the Z-acrylate

Now, the Z-isomer could be directly converted into α,β-unsaturated lactone **151** by treatment with concentrated hydrochloric acid in ethanol. (Scheme 34) To most effectively block one face of the butenolide, the bulky TBDPS group was chosen to protect the primary alcohol. This turned out to be more difficult than expected – according to a literature report,[58] **151** was prone to racemisation upon TBDPS-protection using NEt$_3$ as a base. The equilibrium of the alcoholate **154** with resonance-stabilised furan derivative **155** was thought to be responsible for this. To circumvent this problem, the authors used NH$_4$NO$_3$ as an alternative base; in our hands, this procedure led to erratic results, also accompanied by some racemisation. However, silylation using imidazole and TBDPSCl in a 1:1 ratio at 0 °C gave the desired product **152** in excellent yield with only minimal loss of optical purity (≥ 95 % ee). As an alternative, also the protection of **151** as the equally bulky trityl ether **153** using trityl chloride in pyridine worked fine (as long as no DMAP was used – in this case, partial dimerisation due to a concurrent Baeylis-Hilman reaction was observed). Regrettably, the enantiomeric purity of the product could not be assured due to UV-active impurities that hindered a clear peak assignment.

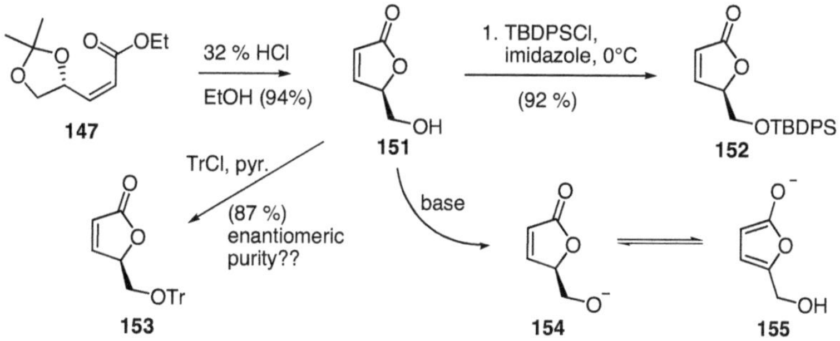

Scheme 34: Synthesis of the TBDPS-protected chiral butenolide **152**

The minor *E*-isomer could also be converted to **151** using the following sequence:[59] Treatment of **148** with thiophenol and Hünig's base, followed by cyclization under

---

[58] Fazio, F.; Schneider, M. P. *Tetrahedron: Asymmetry* **2000**, *11*, 1869 – 1876.
[59] Takano, S.; Kurotaki, A.; Takahashi, M.; Ogasawara, K. *Synthesis* **1986**, 403 – 406.

acidic conditions led to the lacton **156a** as an 3:2 epimeric mixture. (Scheme 35) Oxidation to the sulfoxide **157a** was achieved using NaIO$_4$, followed by elimination under thermal conditions to yield butenolide **158**. However, a different outcome was observed when Tr-protected **156b** was used (to get safe access to **153**). (Scheme 35) After oxidation to the sulfoxide **157b**, no elimination to the α,β-unsaturated lactone was observed. This indicates that the latter reaction is not a thermally induced Cope-elimination using CaCO$_3$ as acid scavenger, but rather an E$_{1CB}$ – reaction, initiated by an intramolecular deprotonation induced by the primary alcohol, which is not possible when this alcohol is protected as the trityl ether.

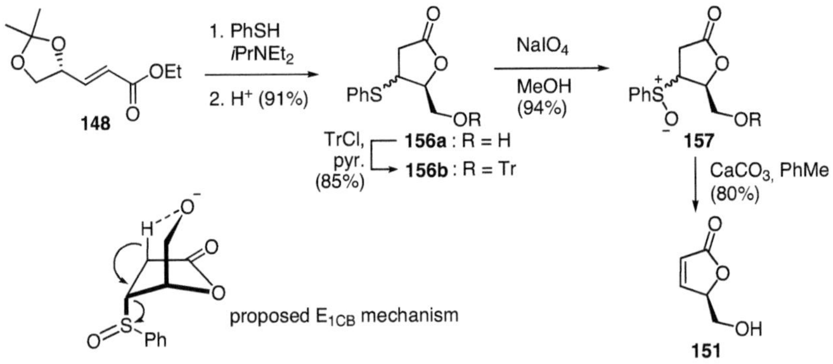

Scheme 35: Conversion of the *E*-isomer to the chiral butenolide **151**

To obtain the desired *trans*-substituted lactone, the following cuprate 1,4-addition should be directed by the steric hindrance of the bulky protecting group to occur in high selectivity from the opposite side. (Scheme 36) Despite discouraging reports in the literature,[60] previous investigations by C. Pilger (see section 3.3) showed that such a 1,4-addition of a vinyl metal species to give **158** was indeed possible, when vinyl magnesium chloride and catalytic amounts (7 %) of CuCl and LiCl in a 1:2 ratio were employed. (Scheme 36) With this reaction, the orthogonally addressable, disubstituted γ-lactone **158** could be accessed.

---

[60] Nilsson, K.; Ullenius, C. *Tetrahedron* **1994**, *50*, 13173 – 13180.

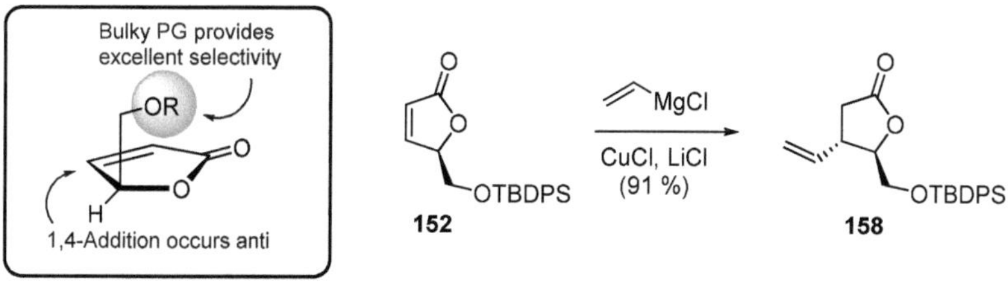

Scheme 36: Cu-catalysed 1,4-addition to yield **158**

6.3   Efforts towards a C-2 functionalization

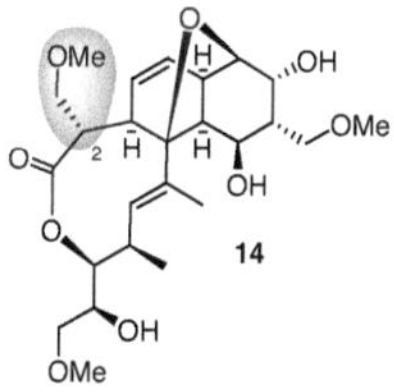

To incorporate as much functionality at an early stage as possible, it was thought that the C-2 methoxymethylene group of the final molecule could already be introduced at this stage, especially as C. Pilger had shown that an *in situ*-aldol reaction of butenolide **93** with acrolein was possible (see section 3.3).

First attempts to directly add MOMCl to the 1,4-addition reaction mixture, failed to give any conversion to α-substituted product **159**; addition of tetrabutyl ammonium iodide (TBAI) did not give any improvement. (Scheme 37) To silence concerns about the lower nucleophilicity of the copper enolate, 1,4-addition product **158** was deprotonated with lithium diisopropylamide (LDA) or lithium hexamethyldisilazide (LiHMDS) and MOMCl was added (also using TBAI and HMPA as additives). However, no conversion could be detected here as well.

As alkylations with MOMCl were only rarely reported in the literature, we decided to try a stepwise synthesis. (Scheme 37) Addition of formaldehyde to the enolate, followed by methylation of the primary alcohol should lead to the related lactone **160**. Unfortunately, neither introduction of formaldehyde gas, generated by careful pyrolysis of *para*-formaldehyde, into the 1,4-addition reaction mixtures nor addition to the enolate created with LDA, resulted in the formation of primary alcohol **160**.

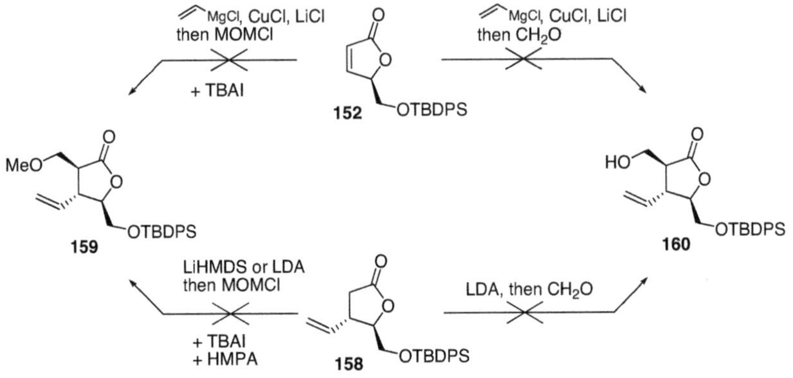

Scheme 37: Attempts to install the C-2 methoxy-methylen substituent

Serendipitously, an alternative way for C-2 functionalisation was found: In an attempt to evaluate different protecting groups (especially regarding the conditions necessary for cleavage), butenolide pivaloate ester **161** was prepared. (Scheme 38) When this Michael acceptor was subjected to the 1,4 – addition as described previously, instead of the desired 1,4-addition product a more polar product bearing a free hydroxy-function was isolated. NMR-analysis revealed it to be the C-2 acylated compound **162**. This surprising reaction can be explained by an intramolecular attack of the enolate derived from the 1,4 addition on the pivaloate – this migration can be arranged in 6-membered transition state **162A**, which seems quite favourable.

We thought that this side-reaction could be used to our advantage, when instead of the pivaloate a formiate was used, this migration would directly install the desired C-O functionality. (Scheme 38) Introduction of the formiate **163** was possible by direct esterification of **151** with formic acid; subjection of this molecule to the 1,4-addition conditions resulted in the formation of a new, polar product, which turned out to be completely insoluble in CDCl$_3$ – NMR analysis had to be performed in MeOH-d$_4$. As no signal corresponding to a proton in α-position to the carbonyl was present, the enolic structure **164** was proposed for this product. Unfortunately, neither protection of the free alcohol was possible nor a reduction of the aldehyde (probably due to full enolisation) was possible, so this strategy had to be abandoned.

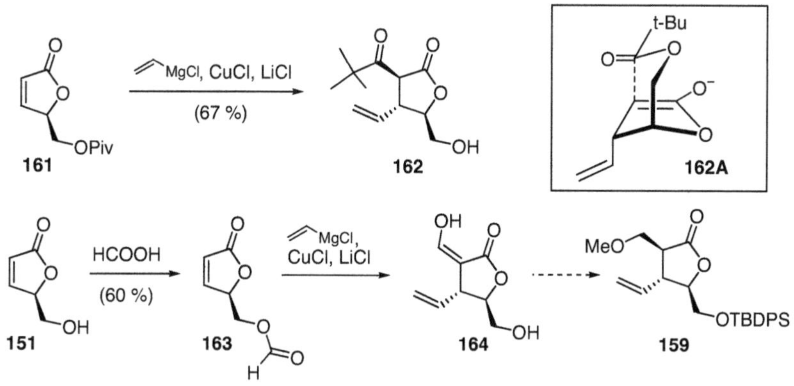

Scheme 38: Serendipitious C-2 acylation and its exploitation

C-2 functionalisation could also be achieved *via* a Mukaiyama-aldol reaction. However, formation of the TMS-enol ether **165** failed; (Scheme 39) Another possibility was to functionalise C-2 prior to the 1,4-addition using a Baeylis-Hilman reaction;[61] but again, upon treatment of **153** with an aqueous solution of formaldehyde and catalytic amounts of DMAP,[62] no conversion to **166** was observed.

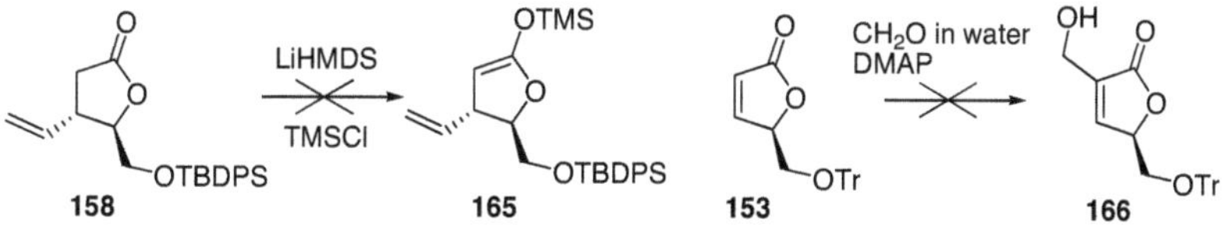

Scheme 39: Further C-2 functionalisation attempts

As all alkylation and acylation attempts of lactones like **158** had failed, strategies to construct an already C-2 functionalised α,β-unsaturated, chiral γ-lactone were

---

[61] For a review on reactions for this purpose, see: Rezgui, F.; Amrib, H.; El Gaïed, M. M. *Tetrahedron* **2003**, *59*, 1369 – 1380.

[62] Rezgui, F.; El Gaïed, M. M. *Tetrahedron Lett.* **1998**, *39*, 5965 – 5966.

investigated. If an α-substituted butenolide **167** is desired, it can be obtained from a saturated lactone **168**, where R could be both the –CH$_2$OMe group or a leaving group to generate the double bond – and both functional group can be introduced *vice versa* by alkylation of the lactone enolate. (Scheme 40) The ring-opened γ–hydroxy-acid derivative **169**, which can be seen as a homoaldol product, can be synthesised by reaction of the dianion of an accordingly functionalised carboxylic acid **170** with protected, enantiopure glycidol **171**.

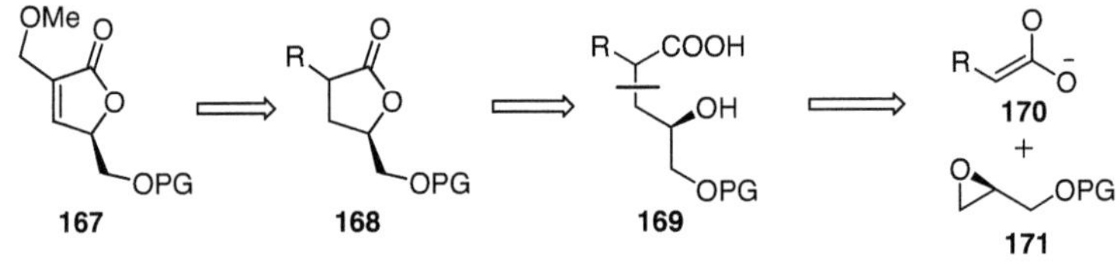

Scheme 40: Alternative butenolide synthesis

As a test substrate, propionic acid **170a** was treated with 2 equivalents of LDA in THF, upon which the formation of a white precipitate was visible. (Scheme 41) In an unoptimised procedure, addition of PMB-protected glycidol **171a** led to the formation of the C-2 methylated γ-lactone **173** – the intermediate γ-hydroxy acid **172** could not be observed.[63] This C-2 epimeric mixture could then be deprotonated with LDA and alkylated with phenylselenyl bromide,[64] although only with low conversion. Oxidation of **174** and elimination of the resulting selenoxide gave the desired α-substituted butenolide **175**. With these positive results, we tried the same sequence using β-methoxypropionic acid **170b** and trityl protected glycidol **171b**. (Scheme 41) However, in this case, no formation of **176** was observed, probably due to decomposition of the high energy acid dianion into methoxide and acrylate.

---

[63] Takano, S.; Tanaka, M.; Seo, K.; Dirama, M.; Ogasawara, K. *J. Org. Chem.* **1985**, *50*, 931 – 936.
[64] Mori, K.; Khlebnikov, V. *Liebigs Ann. Chem.* **1993**, 77 – 82.

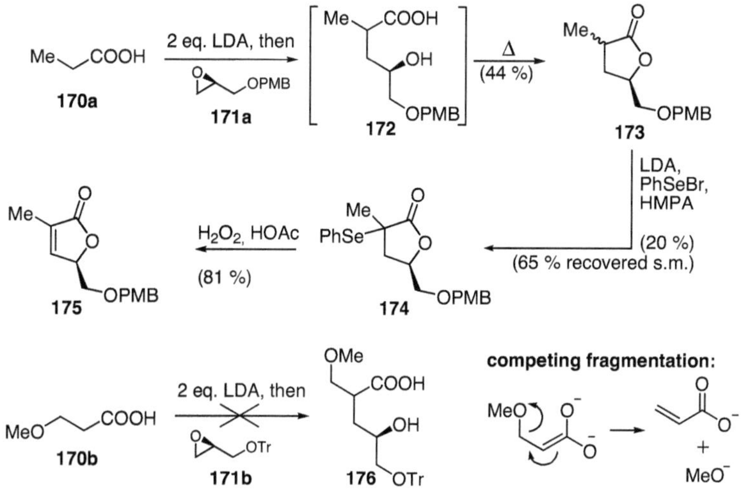

Scheme 41: Butenolide formation by addition of a leaving group after cyclisation

Alternatively, also the leaving group can be introduced with the dianion. This was achieved by deprotonation of phenylthioacetic acid (**170c**) and treatment with trityl-protected glycidol **171b**. (Scheme 42) Phenylthio lactone **178** could be accessed in good yields as a nearly 50:50 epimeric mixture, which was then treated with LDA and gaseous formaldehyde to give epimeric **179**,[65] now as a 2:1 mixture in preference of the *anti*-diol. Oxidation of **179** to the sulfoxide was possible using *m*CPBA – the crude product could then be eliminated to the α,β-unsaturated lactone **180** in refluxing toluene in the presence of CaCO$_3$.

---

[65] Calderon, A.; de March, P.; El Arrad, M.; Font, J. *Tetrahedron* **1994**, *50*, 4201 – 4214.

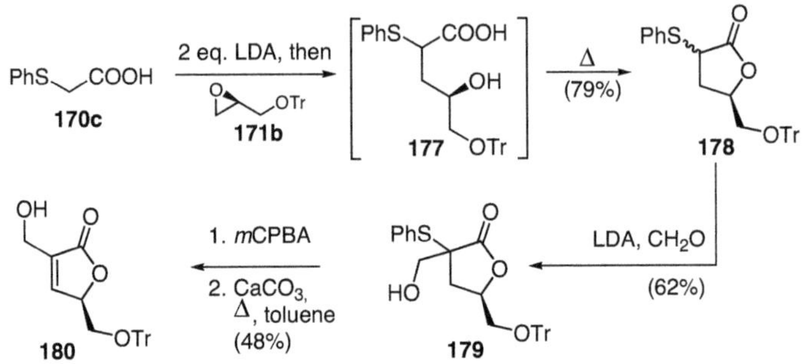

Scheme 42: Butenolide formation by addition of the CH$_2$OH-group after cyclisation

These results suggest, that a fast synthesis of an α-substituted butenolide is indeed possible. However, at the time of these investigations, concerns about the feasibility of the IMDA reaction (*vide infra*) were growing, so all efforts were concentrated on this, and the 1,4-addition to **180** was never tried.

6.4   *E,Z*-Diene building block

6.4.1 Wittig-approach

As we wanted to install the diene moiety in one piece, we considered the fastest way to construct **181** would be using a Wittig-reaction of *E,Z*-double-allylic Wittig reagent **183** with aldehyde **182**. (Scheme 43) Following literature precedence, the configuration of the newly formed double bond should be predominantly Z when bulky aldehydes are treated with allylic Wittig reagents.[66] However, it had to be proven experimentally, if aldehyde **182** could really be regarded as bulky as *t*-BuCHO, which was studied in literature.

---

[66] Tamura, R.; Saegusa, K.; Kakihana, M.; Oda, D. *J. Org. Chem.* **1988**, *53*, 2723-2728.

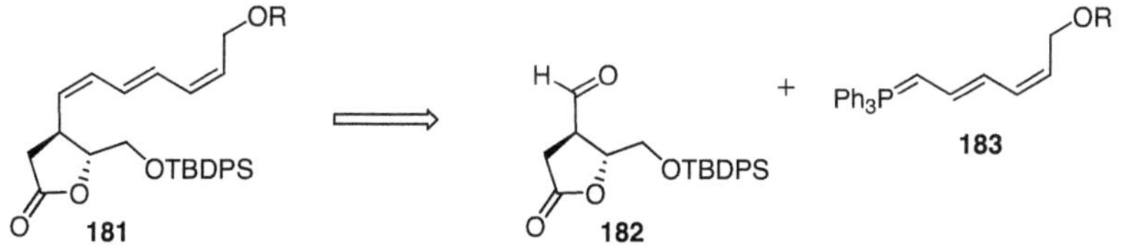

Scheme 43: Triene construction *via* Wittig-olefination

Based on previously gained knowledge about the "do's and don't's" in the synthesis of the *E,Z*-diene, a cross-coupling reaction was thought to be the most reliable method to assemble the diene in a geometrically defined way. However, we thought to construct the Z-double bond part in a different manner. To avoid volatility problems, the methyl ether was replaced by a heavier protecting group, which was also necessary as our new approach necessitated this alcohol to be deprotected at a later stage. Furthermore, to avoid the finicky diimide reduction, the Z-double bond should be constructed in an olefination reaction instead.

Protection of inexpensive *E*-butene-1,4-diole (**184**) as the bis-THP ether **185** could be achieved quantitatively; ozonolysis yielded glycol aldehyde **186**, which was then olefinated using Stork's conditions[67] to give the Z-vinyliodide **187** in good (83 %), but not always reproducible yields. (Scheme 44)

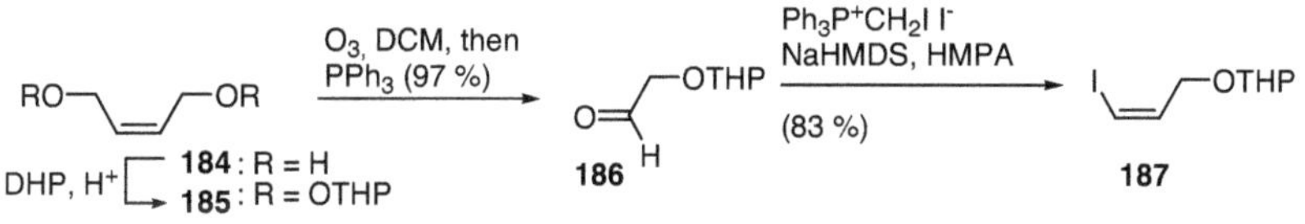

Scheme 44: 2nd Z-Iodo allylic alcohol synthesis

The required coupling partner was synthesised from propargyl alcohol by hydrostannylation under radical conditions to yield **188** (separable from the other two regioisomers by column chromatography).[68] (Scheme 45) Stille coupling with **187**

---

[67] Stork, G.; Zhao, K. *Tetrahedron Lett.* **1989**, *30*, 2173 – 2174.
[68] Jung, M. D.; Light, L. A. *Tetrahedron Lett.* **1982**, *23*, 3851 – 3854.

under standard conditions[69] yielded the *E,Z*-bis-allylic alcohol **189**. Formation of the bis-allylic bromide **190** was possible by formation of the mesylate and *in situ* S_N2 reaction at 0 °C with sodium bromide. This reaction, originally reported for the replacement of mesylate by chlorine using lithium chloride,[70] turned out to be tricky. When this reaction was scaled up, probably due to less efficient cooling, chloride ions from the mesyl chloride also took part in the reaction and a nearly 1:1 mixture of allylic bromide and chloride was formed. **190** could be converted to phosphonium salt **191** by treatment with PPh₃ in acetonitrile at r.t.[71] Although formed quantitatively, purification of this hygroscopic phosphonium salt was quite difficult – after several recrystallization attempts, the crude material was used.

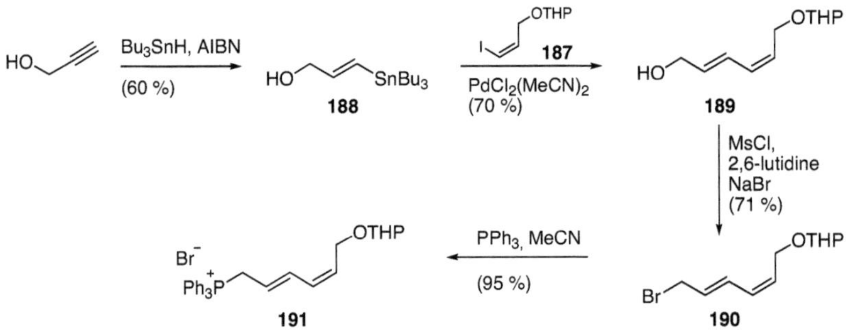

Scheme 45: Construction of the double-allylic phosphonium salt

Ozonolysis of **158** at -78 °C followed by work-up with PPh₃ afforded the aldehyde **182** in good yields. (Scheme 46) Upon deprotonation of phosphonium salt **191** with a variety of lithium bases and addition of **182**, only minor amounts of **192** could be isolated; the best results were achieved when *n*-BuLi was used as a base. In each case, mixtures of isomers were obtained, visible both by TLC and ¹H NMR analysis. Due to

---

[69] a) Stille, J. K.; Simpson, J. H. *J. Am. Chem. Soc.* **1987**, *109*, 2138 – 2152; b) Stille, J. K. *Angew. Chem. Int. Ed.* **1986**, *25*, 508-524; *Angew. Chem.* **1986**, *98*, 504 – 519.
[70] Collington, E. W.; Meyers, A. I. *J. Org. Chem.* **1971**, *36*, 3044 – 3045.
[71] Siegel, K.; Brückner, R. *Chem. Eur. J.* **1998**, *4*, 1116 – 1122.

this disappointing outcome, the Wittig-route to construct the Z,E,Z-triene was abandoned.

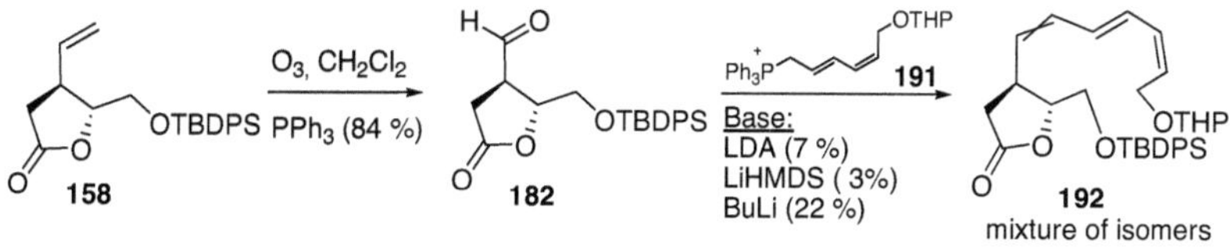

Scheme 46: Wittig-olefination attempts

6.4.2 Cross-coupling approach

The previous experiments had shown that a convergent assembly of the Z,E,Z-triene using olefination chemistry was not possible. An alternative strategy for the construction of triene **181** uses a cross-coupling reaction between C-5 and C-6. Again, it seemed easier to have the iodide located on the Z-configured lactone building block **193** and an E-vinyl stannane fragment **194**. (Scheme 47)

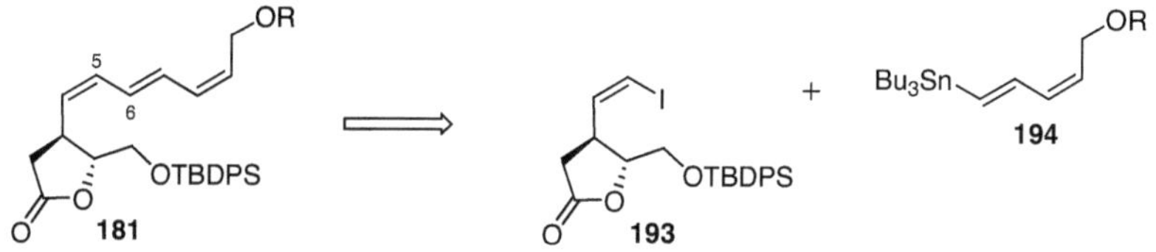

Scheme 47: Triene construction *via* Stille-coupling of Z-vinyl iodide **193** and vinyl stannane **194**

Fortunately, at the time this synthesis was initiated, an applicable synthesis for an *E,Z*-tributylstannyl dienoate had appeared in the literature.[72] Therein, the authors stated that the Z-selective formation of *E,Z*-tributylstannyl dienoate **198** was only possible when *E*-tributylstannyl-propenal **195** (available from **188** by MnO_2 oxidation) was treated with the Still-Gennari phosphonate **196** under the optimised conditions.[73] (Scheme 48) Here, the normally equally reliable Ando phosphonate **197** produced nearly a 1:1 mixture.

---

[72] Franci, X.; Martina, S. L. X.; McGrady, J. E.; Webb, M. R.; Donald, C.; Taylor, R. J. K. *Tetrahedron Lett.* **2003**, *44* 7735 – 7740.

[73] Still, W. C.; Gennari, C. *Tetrahedron Lett.* **1983**, *24*, 4405 – 4408.

These findings shed a light on the peculiar situation present in vinyl-tin compounds: Normally, the addition of a phosphonate stabilised enolate to an aldehyde forms reversibly both the *erythro* (**199**) and *threo* (**200**) adducts (leading to Z and E olefins, respectively), where the *erythro* adduct **199** is lower in energy. However, the rate determining step is the formation of the oxophosphetane, where the *threo* adduct **200** would lead to a lower energy transition state **200A**, giving the *E*-olefin. This means, in the case of 'normal' phosphonates the relative energy of the oxophosphetane intermediate determines the outcome of the reaction. When using an Ando (**197**) or Still-Gennari phosphonate (**196**), formation of the aldol-adduct is irreversible, so the preferentially formed *erythro* adduct **199** leads to the Z-olefin *via* oxophosphetane **199A**.[74] This quite general rule gets disrupted when the electrophile bears a tributyltin substituent, which seems to destabilise *erythro*-adduct **199** and make its formation either unfavourable or reversible. Interestingly, this is only the case when using the Ando phosphonate **197**, but not the Still-Gennari phosphonate **198** – maybe due to a destabilizing interaction between the trifluoromethyl groups and the tributyltin residue, which in turn favours again **199**.

---

[74] Ando, K. *J. Org. Chem* **1999**, 6815 – 6821.

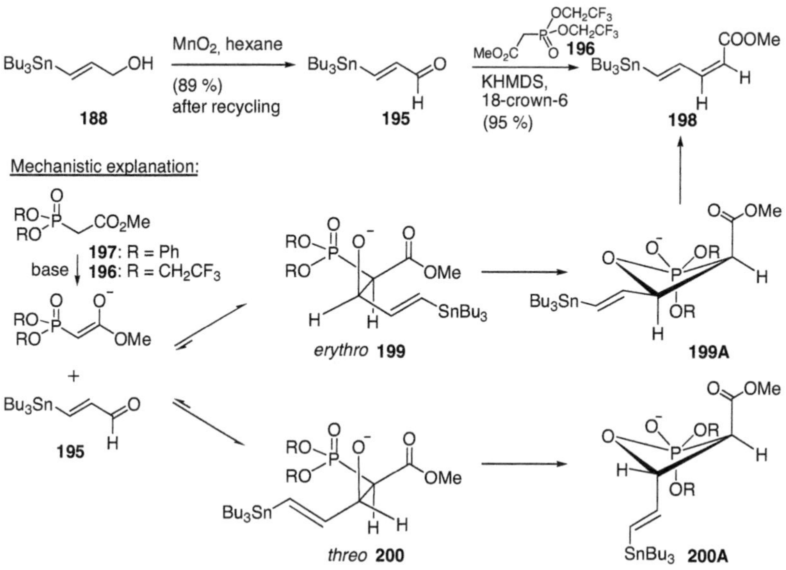

Scheme 48: Construction of the E,Z-tributylstannyl dienoate[72] and its rationale

After the Z-selective olefination, the dienoate could be reduced with DIBAl-H to give allylic alcohol 201 without concomitant 1,4 – or 1,6 reduction in excellent yields. (Scheme 49) However, the previously anticipated THP protecting group could not be installed on the primary allylic alcohol, as proto-destannylation occurred under acid-catalysed acetalisation conditions. Instead, PMB was chosen as alternative, as it could be installed under basic conditions, and its removal should be unproblematic at a later state.

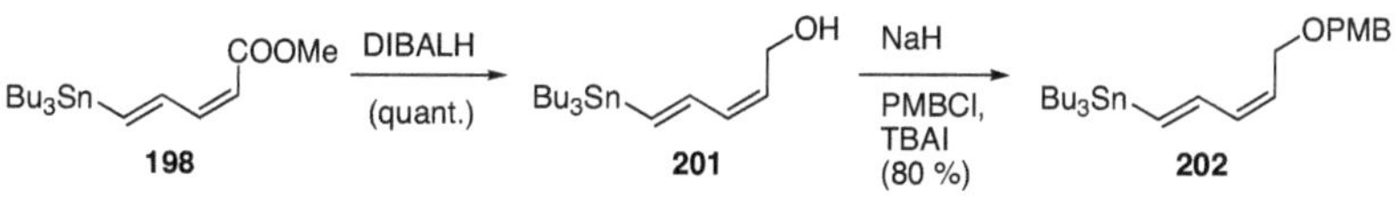

Scheme 49: Synthesis of E,Z-dienyl stannyl alcohol 202 building block

Having now the desired vinyl-tin compound 202 in hand, we next turned to synthesise the coupling partner – for highest activity in the coupling reaction, a Z-vinyl iodide was most desirable. By applying the Stork conditions to aldehyde 182 (see section 6.4.1), Z-iodovinyl 193 was produced, although this time in considerably lower yield. (Scheme

50) Nevertheless, vinyl iodide **193** was coupled with dienyl stannane **202** to give triene **203**.

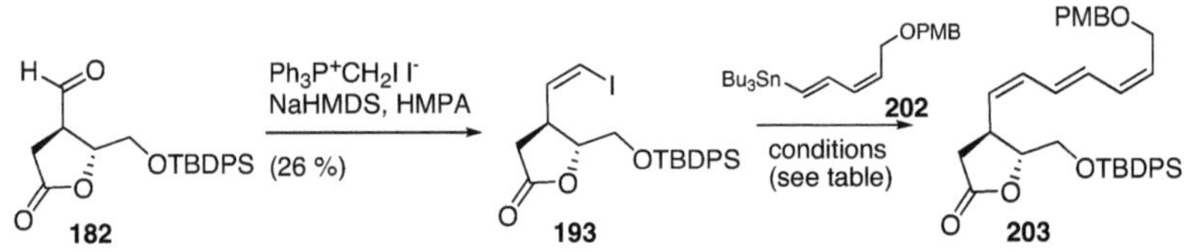

Scheme 50: Z,E,Z-triene synthesis *via* Stille-coupling

Although previously applied conditions led only to meagre results, the use of more sophisticated methods applied in recent total syntheses projects[75,76] led to formation of the desired triene **203** in good yields. (Table 1)

Ref.	Catalyst	Additives	T (°C)	Solvent	Yield
69	$PdCl_2(CH_3CN)_2$	-	r.t.	DMF	44 %
75	$Pd(PPh_3)_4$	CuCl	r.t.	DMF	56 %
76	$Pd_2(dba)_3$	$P(2\text{-furyl})_3$, CuI	r.t.	DMF	68 %
76	$Pd_2(dba)_3$	$P(2\text{-furyl})_3$, CuI	45 °C	DMSO	84 %

Table 1: Coupling conditions to install the triene

Removal of the TBDPS – group was less successful – all attempts to cleave the silyl group in **203** using either HF·pyridine, TBAF in THF solutions or crystalline TBAF·3H$_2$O (using THF, or benzene as solvents, adding 4 Å molecular sieves) only led to low and unreproducible yields of **204**. (Scheme 51)

---

[75] Munakata, R.; Katakai, H.; Ueki, T.; Kurosaka, J.; Takao, K.-i.; Tadano, K.-i. *J. Am. Chem. Soc.* **2003**, *125*, 14722 – 14723.

[76] Kadota, I.; Takamura, H.; Sato, K.; Ohno, A.; Matsuda, K.; Satake, M.; Yamamoto, Y. *J. Am. Chem. Soc.* **2003**, *125*, 11893 – 11899.

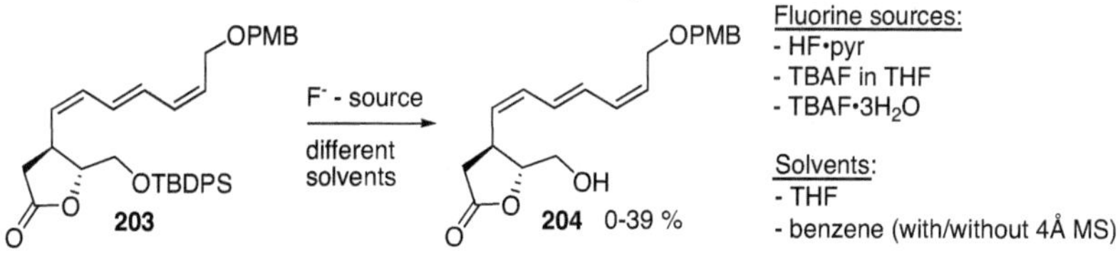

Scheme 51: Deprotection attempts

As TBAF is known to be a variably basic reagent (depending on the water content), the suspicion arose that the lactone moiety in **203** might be sensitive to base – such susceptibility would also explain the low and unreproducible yields in the formation of **193** *via* Wittig olefination, another naturally basic reaction. As this basic reaction was quite fundamental in this synthetic sequence, a protection of the base-sensitive lactone was deemed to be a good method to solve both problems.

6.4.3 Cross-coupling / lactol approach

A useful way to protect the lactone tether was to convert it to the methyl lactol. Lactone **158** was reduced to the lactol **205** by DIBAl-H reduction, followed by treatment of the crude product with $BF_3 \cdot OEt_2$ in methanol to give methyl acetal **206**. (Scheme 52) This reaction was carried out at 0 °C, and produced both epimers of the acetal **206** in a 55:45 ratio. The reaction sequence was continued with this epimeric mixture, until both epimers could be separated at a later stage. Then, these epimers were used individually – although not all reactions were performed with both acetals. After ozonolysis, the olefination of aldehyde **207** now worked with much better, but still not always reproducible yields. The produced epimeric vinyl iodides **208a** and **208b** could be separated by column chromatography.

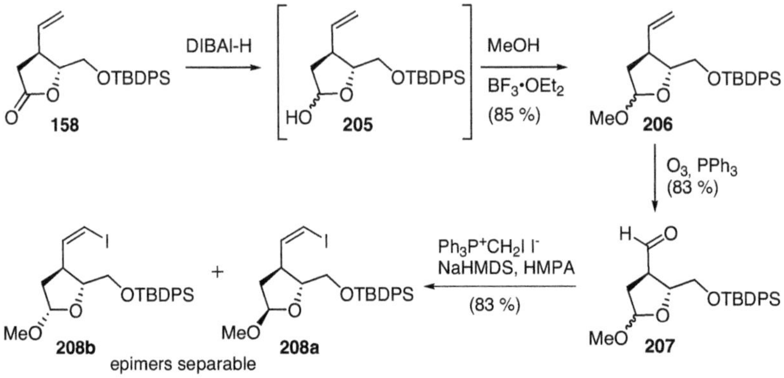

Scheme 52: Use of a methyl lactol as protection of the lactone

As the relatively sensitive triene should be introduced as late as possible into the molecule, we first wanted to convert the primary alcohol into the dienophile. However, when TBDPS-deprotection of Z-vinyliodide **208b** was executed using TBAF, only alkyne **210** could be isolated! (Scheme 53) This surprising finding can be explained by an E$_2$-elimination taking place right after deprotection, with the newly formed alkoxide acting as an internal base, attacking the β-vinyl hydrogen through a 6-membered ring transition state **209**.

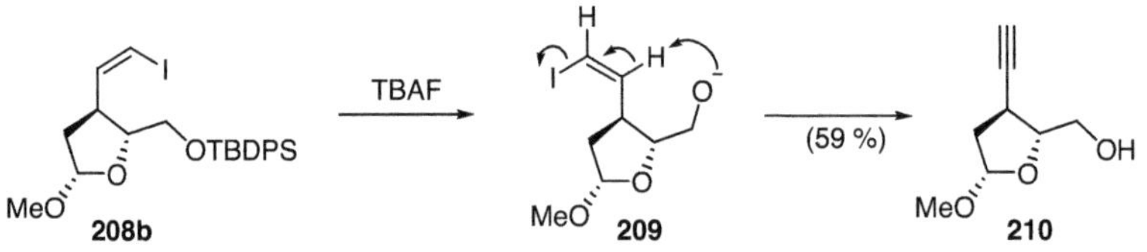

Scheme 53: Concomitant E$_2$-elimination with the deprotection

To circumvent this reaction, we had to go back to methyl lactol **206** and do without the separation of the epimers on the stage of the Z-vinyl iodides **208**. TBAF-deprotection of **206** yielded volatile primary alcohol **211** in excellent yields, (Scheme 54) which could be oxidised to the aldehyde **212**, using either the Swern conditions or Dess-Martin periodinane. Formation of the desired Z-α,β-unsaturated ester **213** was possible using the Still-Gennari reagent, although this time with some concomitant formation of the undesired E-isomer. **213** now contained two double bonds, one electron-rich (vinyl group) and one electron-poor (enoate group), which could be therefore easily differentiated. Catalytic osmylation followed by periodate cleavage gave aldehyde **214**, which was then transformed into Z-vinyliodide **215**, although again in poor yields. Apparently also the enoate acts as a competing electrophile or enolizable position. As working with an epimeric mixture made the NMR analysis quite difficult and the bad yield in this last step was not acceptable, we had to find a better way to access **216**.

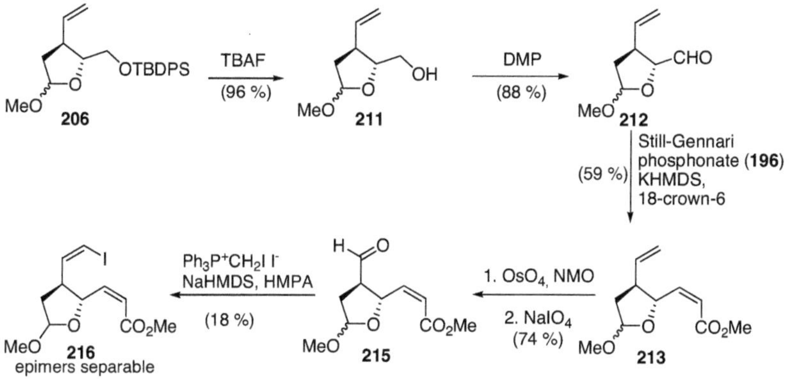

Scheme 54: Synthesis of the Z-vinyl iodide coupling partner

Such a solution was found when instead of using TBAF, the deprotection of **208b** was done with HF·pyridine, which produced the desired primary alcohol **216b**. (Scheme 55) Still, special care had to be taken upon column chromatography, as the free hydroxy group neighboring the methyl lactol seemed to facilitate its decomposition on silica gel when less than 5 % NEt$_3$ was added to the eluent. Oxidation to the corresponding aldehyde **217b** proved to be difficult, and could at the time be only achieved in 53 % yield using the Parikh-Doering protocol (SO$_3$·pyridine/DMSO).[77] Due to bad experience with the Still-Gennari phosphonate (*cf.* Scheme 54) the Ando phosphonate **197** was used instead and led to comparably clean product **218** in good yield. Coupling of Z-

---

[77] For a review on DMSO oxidations, see: Tidwell, T.T. *Synthesis* **1990**, 857 – 870.

vinyliode **218** with stannanes **201** and **202** could be achieved under previously optimised conditions (see Table 1) to give tetraenes **219** and **220**.

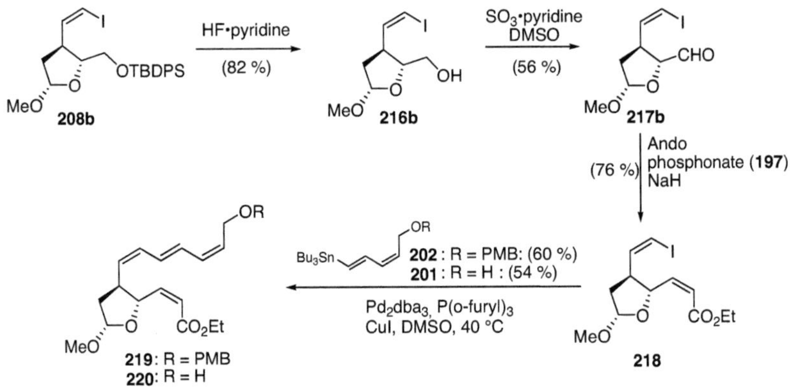

Scheme 55: Synthesis of Z,E,Z,Z-tetraenes **219** and **220**

Initially, the thermal Diels-Alder reaction of PMB-protected tetraene **219** was examined. As no conversion was visible when **219** was heated to reflux in toluene for 6 h, the solution was transferred to a sealed tube and heating was continued at 190 °C bath temperature for 24 h. (Scheme 56) Again, as TLC analysis showed no formation of a new product, the reaction was stopped and purified by column chromatography. Besides a small amount of isomerised terminal Z-allylic alcohol, comparison with the crude spectrum showed no further compounds present.

We therefore reasoned that under the reaction conditions, diene and dienophile entropically do not come close enough to undergo a Diels-Alder cyclization. It seemed logical, that the free terminal alcohol in **220** could be used to bring diene and dienophile closer together when transformed into its metal alkoxide. Such a counterion could act as an 'internal' Lewis-Acid, to activate the dienophile both thermodynamically (acting as a

Lewis-acid) as well as entropically (by forming a temporary bond to the carbonyl).[78] Such temporary tethers have been realised as silicon,[78b,c] aluminium,[78d] zinc,[78e] boron[78f] and mangnesium[78d] alkoxides. For reasons of simplicity, we chose to transform the allylic alcohol **220** into its magnesium alkoxide. (Scheme 56) After stirring at r.t. for 24 h, the TLC showed a slight change of the R_f-value. A crude NMR spectrum revealed the only visible change to be the multiplicity and shift of the olefinic signals. This product could later be identified as the *Z,E,E,Z*-tetraene **221**. Interestingly, the other 2 *Z*-configured double bonds seemed perfectly stable under the applied conditions.

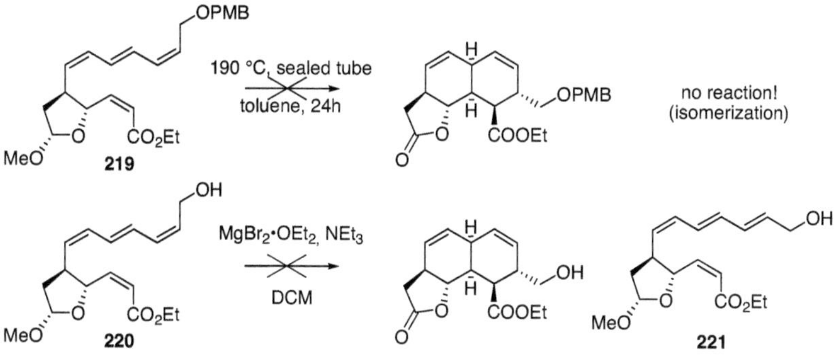

Scheme 56: Assessed IMDA reactions of tetraenes **219** and **220**

As suspicion arose that the *Z*-configured dienophile might impose a steric hindrance, also the use of its *E*-isomer was examined. As the previous approach was hampered by the difficult accessibility of the aldehyde **217**, we thought to circumvent this problem by applying an *in situ* oxidation-olefination method. As aldehydes are often problematic to handle due to instability issues, different solutions have been proposed in the literature

---

[78] a) Barriault, L.; Thomas, J. D. O.; Clément, R. *J. Org. Chem.* **2003**, *68*, 2317 – 2323; b) Sieburth, S.; Fensterbamk, L. *J. Org. Chem.* **1992**, *57*, 5279 – 5281; c) Stork, G.; Chan, T. Y.; Breault, G. A. *J. Am. Chem. Soc.* **1992**, *114*, 7578 – 7579; d) Stork, G.; Chan, T. Y. *J. Am. Chem. Soc.* **1995**, *117*, 6595 – 6596; e) Olsson, R.; Bertozzi, F.; Fredj, T. *Org. Lett.* **2000**, *2*, 1283 – 1286; f) Batey, R. A.; Thadani, A. N.; Lough, A. J. *J. Am. Chem. Soc.* **1999**, *121*, 450 – 451 and references cited therein.

to adress this problem, including one-pot Swern-Wittig combinations[79], PCC-oxidation-Wittig reactions,[80] tandem MnO_2-oxidations/olefinations[81] and a Dess-Martin-Wittig couple.[82] The latter protocol seemed most attractive to us, given that a Dess-Martin oxidation had already worked before to oxidise compound **211**. Indeed, *in situ* oxidation/olefination of **208a** using DMP / benzoic acid / Wittig-reagent gave **222** in good and reproducible yields. (Scheme 57) Coupling to stannane **201** gave the *Z,E,Z,E*-tetraene **223**.

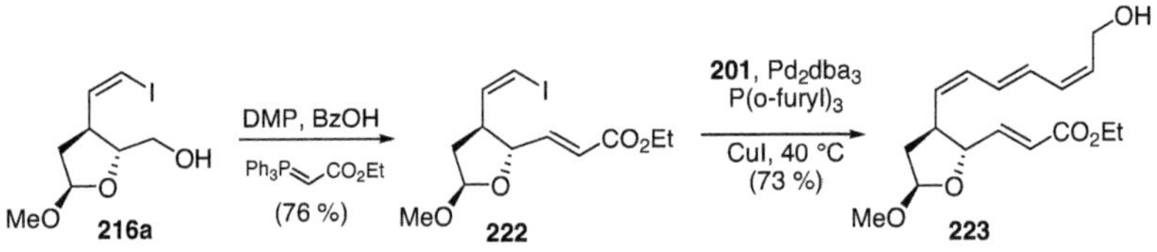

Scheme 57: Synthesis of a *Z,E,Z,E*-tetrane **223**

As the previously assessed Lewis-acid tether was too harsh for the apparently labile *Z*-allylic alcohol, we thought that maybe an intramolecular hydrogen bridge between the free hydroxyl and the carbonyl might provide sufficient complexation to facilitate the cycloaddition. According to findings in the literature,[83] such interactions were strong enough to overrule the *endo*-preference of an IMDA reaction. To further activate the reaction, we thought to decrease the molecular volume by conducting the reaction under high-pressure conditions. Such conditions are reported to be especially benign for 'difficult' Diels-Alder reactions.[84]

---

[79] Ireland, R. E.; Norbeck, D. W. *J. Org. Chem.* **1985**, *50*, 2198 – 2200.
[80] Bressette, A. R.; Glover, L. C. *Synlett*, **2004**, 738 – 740.
[81] a) Blackburn, L.; Xudong, W.; Taylor, R. J. K. *Chem. Commun.* **1999**, 1337 – 1338; b) For a review on this matter, see: Taylor, R. J. K.; Reid, M.; Foot, J.; Raw, S. A. *Acc. Chem. Res.* **2005**, *38*, 851 – 869.
[82] a) Huang, C. C. *J. Labelled Compd. Radiopharm.* **1987**, *24*, 676 – 681; b) Barrett, A. G. M.; Hamprecht, D.; Ohkubo, M. *J. Org. Chem.* **1997**, *62*, 9376 – 9378.
[83] Cayzer, T. N.; Paddon-Row, M. N.; Sherburn, M. S. *Eur. J. Org. Chem.* **2003**, 4059 – 4068.
[84] For a review, see: Matsumoto, K.; Hamana, H.; Iida, H. *Helv. Chim. Acta* **2005**, *88*, 2033 – 2234.

However, when a DCM solution of **223** was subjected to 13 kbar in a hydraulic press for 48 h, a clean isomerisation to the *E*-allylic alcohol was observed, but no formation of a cycloaddition product. (Scheme 58) As isomerisation rather than cycloaddition was observed under all examined reaction conditions, the whole strategy had to be refurbished.

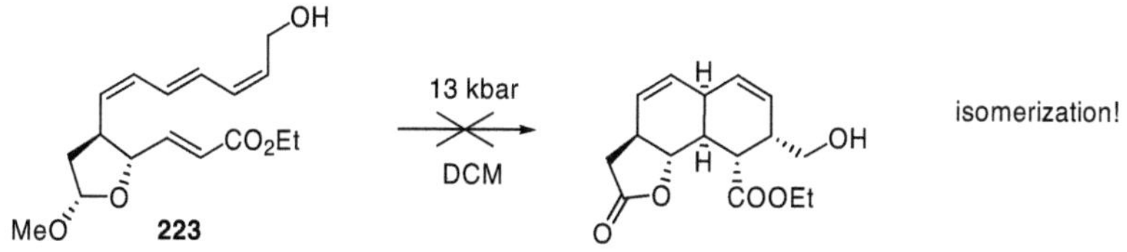

Scheme 58: High-pressure IMDA reaction of **223**

7 3ʳᴰ GENERATION APPROACH

7.1 Concept

In the previous approach to the *cis*-octalin, the tetraene **140** was designed for the easiest Diels-Alder reaction, *i.e.* the dienophile configuration matched with the electronic requirements of the tether to approach the diene in an *endo*-fashion, as depicted in **140A**. (Scheme 59) Such a transition state would have led to product **141**.

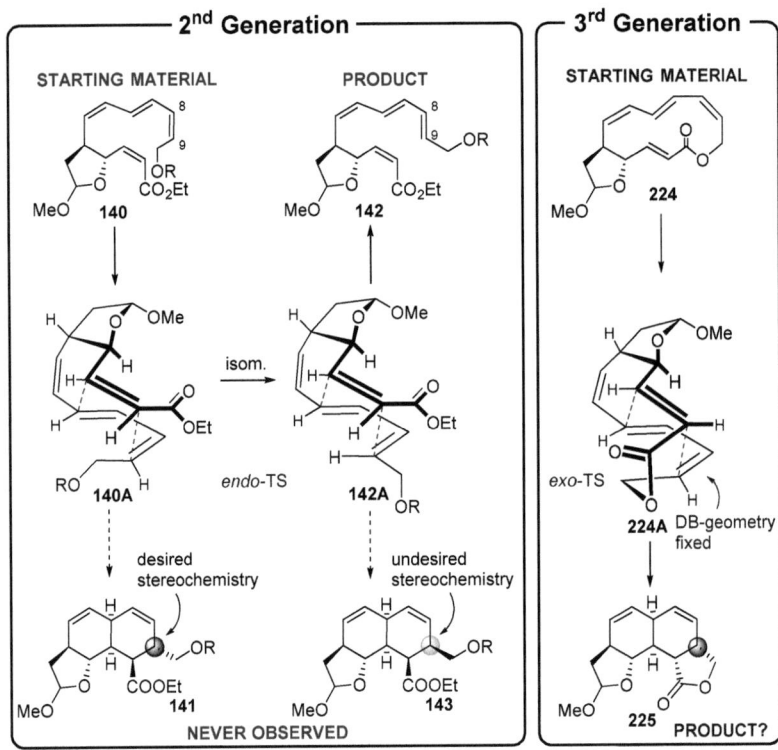

Scheme 59: Comparison 2ⁿᵈ generation *vs.* 3ʳᵈ generation

However, electronic activation turned out not to be the weak point of this system. Instead, it was found that the C-8 – C-9 double bond was prone to isomerisation and under all reaction conditions examined, only isomerisation product **142** was observed.

As this Z,E,E,Z-tetraene, would have given, through transition state **142A**, the wrong stereochemistry at C-10 (*i.e.* hexalin **143**), we had to draw the following conclusion out of these results:

- The absence of any reaction product indicated that diene and dienophile might not get close enough to undergo a Diels-Alder reaction
- As no product is formed even after isomerisation in the presence of a chelating Lewis-acid, either intramolecular complexation is not strong enough, or intermolecular bonding is preferred
- Isomerization of the C-8 – C-9 double bond will continue to be present as long as the *E*-isomer is thermodynamically more stable.

These conclusions suggest that the reaction could be facilitated from an entropic point of view. An obvious method to do so is to use a fixed tether instead of the temporary one, in form of a macrolactone such as **224**. (Scheme 59) In such a macrolactone, only an *E*-configured α,β-unsaturated ester group would point towards the alcohol on the diene. As suggested by transition state 'cartoon' **224A**, this would bring the two reaction partners together quite effectively. If this transannular Diels-Alder (TADA) reaction (now an *exo*-transition state) proceeds along this reaction pathway, *cis*-hexalin **225** should be formed preferably. Still unanswered in this context is the question about the needed higher stability of the Z-allylic alcohol in macrolactone **224**. The construction of a Z,E,E,E-configured macrolactone **229** with a molecular model showed such a system to be very rigid – much more than the corresponding Z,E,Z,E-isomer **224**. To quantifiy this energy difference, the energies of both **224** and **229** were assessed by molecular modelling studies. Calculations of the equilibrium conformers, followed by geometry optimisation using the B3LYP/6-31G* base set proposed an energy difference of 9.2 kcal/mol between the two isomers in favour of the Z,E,Z,E-isomer **229**.

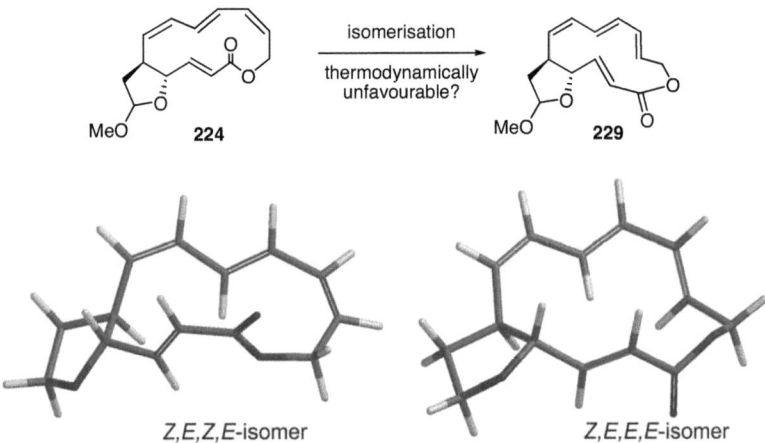

Figure 20: B3LYP-optimised conformers of both isomers **224** and **226**

On this premise, we thought we could also use these *in silico* methods to save valuable experimentation time and get some backup for the concept of stereoinduction in the transannular Diels-Alder (TADA) reaction. As both diene and dienophile are unsymmetrically substituted, four different transition states can be accessed giving rise to four diastereomeric products. These can be clustered in two *endo-* and two *exo-* transition states. (Scheme 60) Inspections of molecular models showed that in the case of the *endo* transition states **224C** and **224D**, due to the double bond geometry, the dienophile is twisted in an antarafacial manner across the diene. Because of these highly strained transition states, visible by the twist of the ester group in the transition state cartoons **224C** and **224D**, we excluded them from further considerations.

Evaluation of the two remaining *exo*-transition states **224A** and **224B**, on the other hand, was not as easy, as both seemed accessible with a molecular model, with a slight preference for **224A**, however without any hard facts to support this proposition. We hoped such facts could be found by a calculation-based assessment of these two competing reaction.

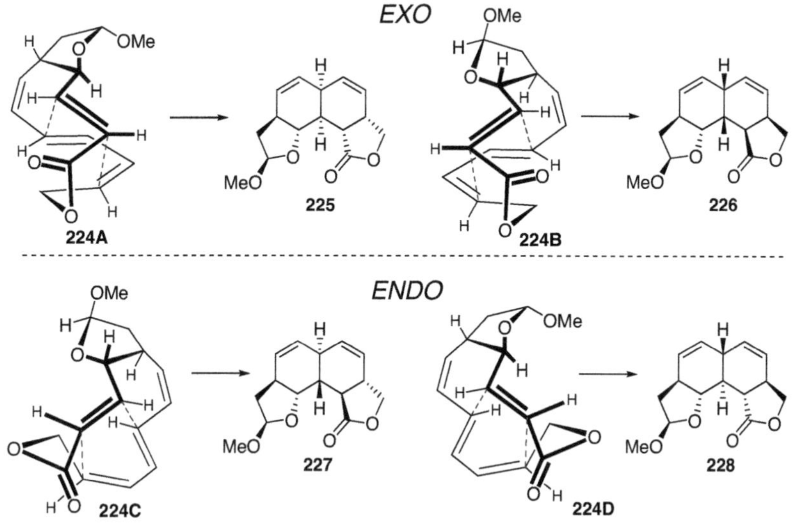

Scheme 60: Possible transition states for the TADA reaction

To be able to compare these two reaction pathways, first both products **225** and **226** were analysed for possible conformers, and after the chemically most promising ones were picked, these were subjected to high quality DFT geometry optimizations (base set B3LYP/6-31G*).[85] The transition states **TS-224A** and **TS-224B** leading to both products were modelled with an equally high level of theory. With both transition and final ground state models in hand, it remained to find the appropriate conformers leading to the corresponding transition states. Those could be found by looking for conformations with similar bond distances as present in the transition state.

Having now calculated all steps of both reaction pathways, a direct comparison of these was possible. (Figure 21) This juxtaposition unequivocally predicted the formation of **225** to be favored. The energy of the conformer **224A*** is 3.6 kcal lower than that of **224B***, which can be explained by the smaller dihedral angle in **224A*** between the two

---

[85] Gaussian 03 Rev. C.02 was used for these calculations.

protons at C7 and C8 (10° opposed to 40°). This results in a better conjugation of the triene, making the energy differences between HOMO and LUMO 0.35 eV smaller. Comparing the transition states, the activation energy for **TS-224A** is 2.2 kcal/mol lower than that of **TS-224B**.

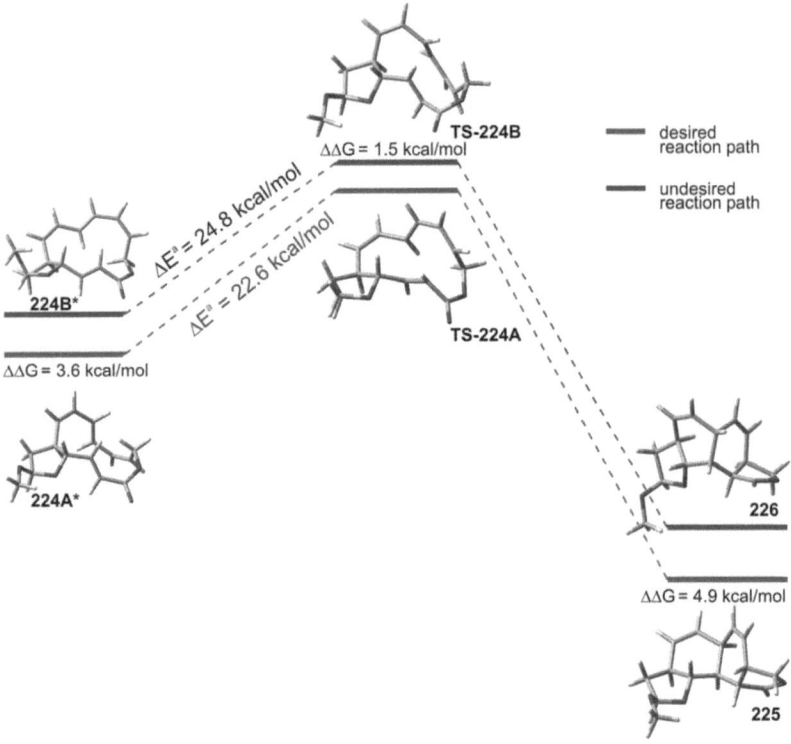

Figure 21: B3LYP-calculated energy profiles of both potential reaction pathways

Furthermore, in **TS-224A** the C-6 – C-11 bonds is 8 nm shorter than in **TS-224B**. (Figure 22) Although both TADA reactions are asynchronous, (seen by the shorter C-9 – C-10 distance), bond formation is further advanced in **TS-224A**. Additionally, the higher dipole momentum of **TS-224B** favors the formation of **225** in an apolar reaction medium.

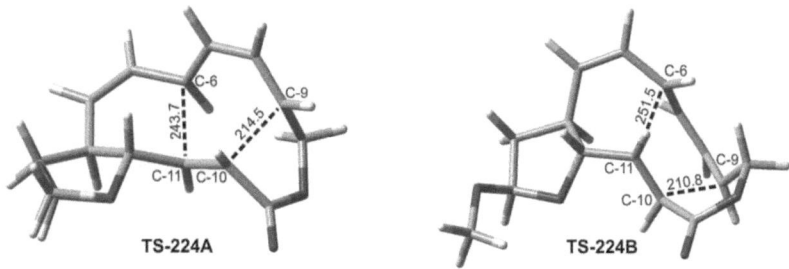

Figure 22: Lengths of forming σ-bonds in the transition states

Finally, it should be noted that the conformation of **225** depicted in Figure 21 is not the direct product of the TADA reaction. (Figure 23) The initially formed product (as can be seen by the antiperiplanar alignment of the protons H-10 and H-11) can undergo a ring flip, releasing about 1 kcal/mol, to produce the ground state conformer, in which in H-10 – H-11 are in a gauche arrangement.

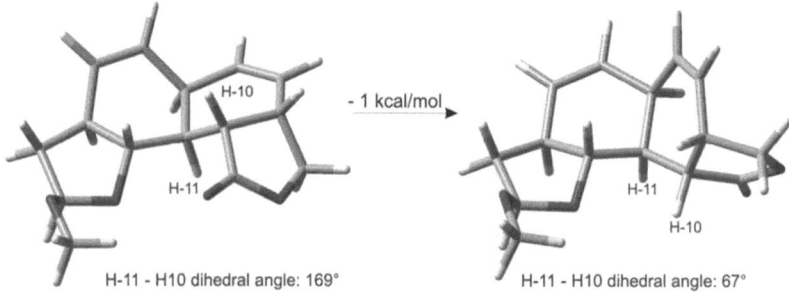

Figure 23: Ring-flip of the Diels-Alder product **225**

7.2   Approaches towards the *Z,E,Z,E*-macrolactone

Motivated by the backing of the DFT-calculations, we started to adopt our synthesis to construct the desired macrolactone. At this time, the synthesis of a related macrolactone was reported in the literature,[86] and the described methodology of this synthesis seemed applicable in our case.

Up to that time, methyl lactol **206** had to be used as 55:45 epimeric mixture. This mixture could be separated at a later stage, but in all subsequent steps characterization had to be done for both epimers. To avoid this, and also to facilitate the pooling of material, another way of acetal formation was sought. It was found that acetal formation under 'basic conditions' (using Ag_2O as base) with MeI as electrophile,[87] gave the α-methyl acetal **206a** in a 95:5 selectivity. (Scheme 61) This greatly enhanced selectivity can be explained mechanistically: under Lewis acidic conditions, oxonium ion **231** is formed, which can be attacked from both α and β face – only little selectivity is observed as the trajectories on both sides are equally shielded by substituents. Under basic conditions, the bulky protecting group on the primary alcohol exerts a repulsive interaction with the acetal anion, thus favouring the formation of **230a**, which, upon attack on the electrophile, leads selectively to the α-methyl acetal **206a**.

---

[86] Suzuki, T.; Usui, K.; Miyake, Y.; Namikoshi, M.; Nakada, M. *Org. Lett.* **2004**, *6*, 553 – 556.

[87] Nicolaou, K. C.; Snyder, S. A.; Huang, X.; Simonsen, K. B.; Koumbis, A. E.; Bigot, A. *J. Am. Chem. Soc.* **2004**, *126*, 10162 – 10173.

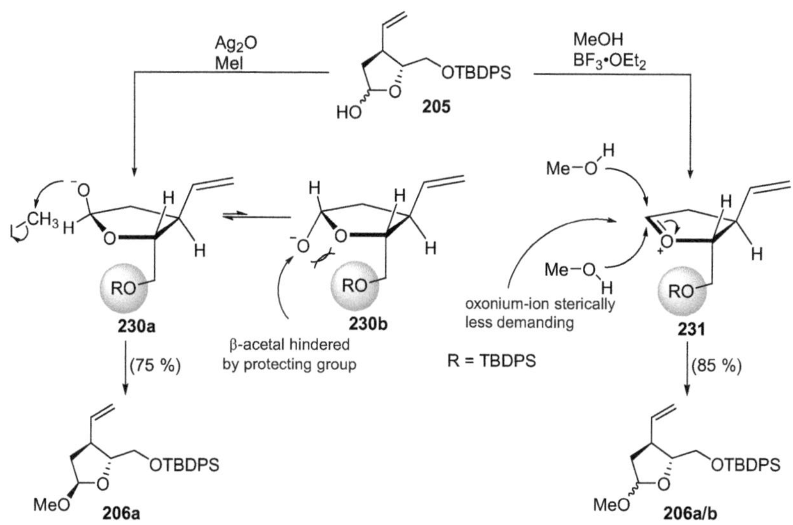

Scheme 61: Selective acetal formation

Being ablo to selectively access one epimer of the lactol, the synthesis of macrolactone **224** was commenced. In the first attempt, we planned to access **224** in an intramolecular Horner-Wadsworth-Emmons (HWE) olefination *via* the *seco*-phosphonate **232**, analogous to the approach reported in the literature.[86] (Scheme 62)

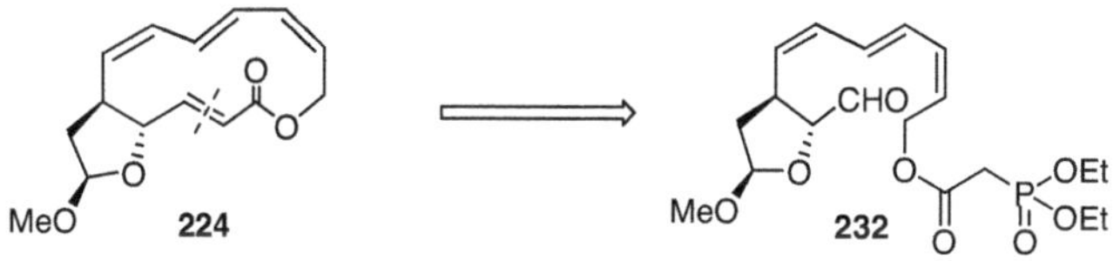

Scheme 62: Intramolecular HWE-olefination

For this approach, stannane **201** had to be adopted: Diethylphosphonoacetic acid (**233**) (synthesised by basic hydrolysis of commercially available triethylphosphonoacetate) was treated with oxalyl chloride at 0 °C to form the acid chloride **234** – after removal of the solvent *in vacuo*, the acid chloride was esterified with the allylic alcohol **201** under DMAP-catalysis, using pyridine as base to form phosphonate ester **235**. (Scheme 63) After coupling with the Z-vinyl iodide **208a**, the resulting phosphonate ester **236** was deprotected to give the polar primary alcohol **237**. However, oxidation of **237** to the aldehyde could not be easily traced, although oxidation seemed to have occurred. In any case, the subsequent intramolecular HWE olefination gave no product.

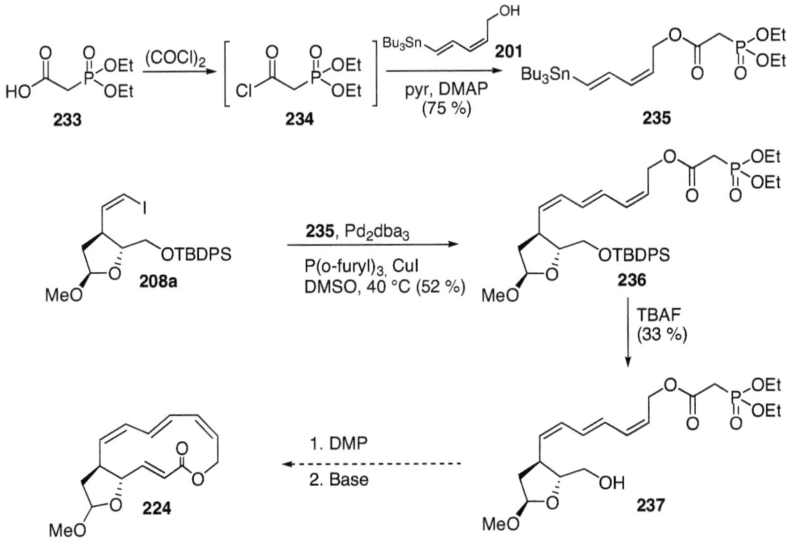

Scheme 63: Attempted synthesis of **224** *via* intramolecular HWE-reaction

7.2.1 C-5 – C-6 Stille coupling approach

Given the already known instability of the C-11 aldehydes (*cf.* chapter 6.4.2, Scheme 55) necessary for the intramolecular HWE-olefination, we quickly realised that this reaction was no good choice for the ring-closure. Instead, by inversion of the reaction sequence, the macrocycle could be closed in an intramolecular Stille-coupling. Such intramolecular couplings are getting increasingly popular in the last years in natural product syntheses.[88]

---

[88] For reviews on this matter, see: a) Duncton, M. A. J.; Pattenden, G. *J. Chem. Soc., Perkin Trans. I* **1999**, 1235 – 1236; b) Nicolaou, K. C.; Bulger, P. G.; Sarlah, D. *Angew. Chem. Int. Ed.* **2005**, *44*, 4442 – 4489.

Compound **208a** was deprotected using NH$_4$F in methanol, which proved to be superior to previous deprotection methods regarding the ease of work-up. (Scheme 64) Subsequent oxidation to aldehyde **217a** was found easiest using a heterogenous reaction in acetonitrile with iodoxyl-benzoic acid (IBX) **238**,[89] which facilitated the work-up tremendously. With this fast and safe oxidation method, the subsequent HWE-olefination with phosphonate **235** (using the especially mild Roush-Masamune protocol[90]) could be achieved in good yield. When *seco*-compound **239** was subjected to various coupling conditions, conversion to several products was observed. The combined efforts are summarised in Table 2.

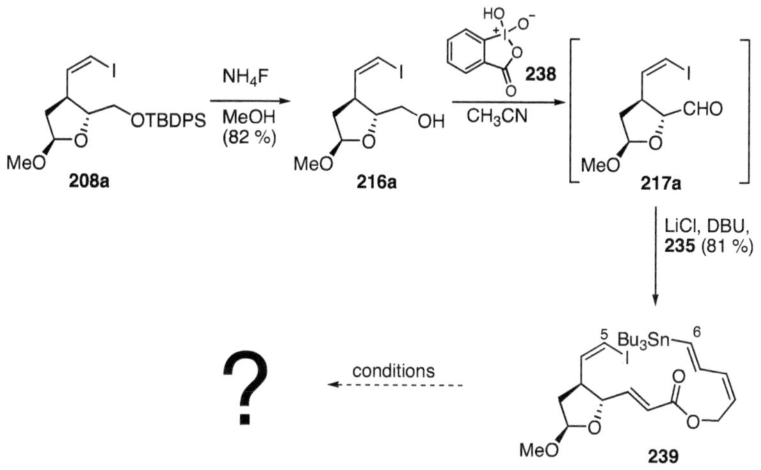

Scheme 64: Attempted sp^2-sp^2 Stille coupling between C-5 and C-6 of **239**

The previously optimised conditions (Table 2, entry 1) led to formation of 2 new products, a major product and a minor one, the latter being further impurified with ligand rests – mass spectroscopy did not show the correct mass peak. Using the conditions applied in Hodgson's periplanone B synthesis[91] (Table 2, entry 2) also did

---

[89] More, J. D.; Finney, N. S. *Org. Lett.* **2002**, *4*, 3001 – 3003.
[90] Masamune, S.; Roush, W. R.; Sakai, T. *Tetrahedron Lett.* **1984**, *25*, 2183 – 2186.

not provide any product. When the Hermann-Beller catalyst **240** (Table 2, entry 3) was used,[92] only destannylated product **241** could be detected. At this time, it seemed promising to switch from Pd-catalysed couplings to a copper mediated cross-coupling using copper-(I)-thiophen-2-carboxylate (CuTC), developed by Liebeskind[93] – on the first attempt, a new product **242**, having the correct mass peak, accompanied by significant amounts of destannylated product **241** could be isolated. Upon further experimentation, formation of this by-product could be suppressed efficiently by using freshly distilled *N*-methyl-2-pyrrolidinone (NMP) under strictly anhydrous conditions. Initial attempts to perform this reaction in high-dilution worked fine, but when raising the concentration slightly, the yield dropped (Table 2, entries 5 and 6) – furthermore, reproducibility was not granted. Therefore we stepped back to Pd-couplings – producing the Pd^0 *in situ* by DIBAl-H reduction (Table 2, entry 7) did not yield any product; however, application of the originally employed conditions using PdCl$_2$(CH$_3$CN)$_2$ as Pd-source gave promising results (Table 2, entries 8-12) – however, yields were again not reproducible, and could not be correlated with dilutions applied (Table 2, entries 10-12).

entry	Ref.	Pd-source	Additives	c [M]	Solvent	T (°C)	Products/remarks
1	76	Pd$_2$(dba)$_3$	P(2-furyl)$_3$, CuI	1*10^{-2}	DMSO	45 °C	2 unidentified prod.
2	91	Pd$_2$(dba)$_3$	AsPh$_3$	5*10^{-3}	NMP	70 °C	unidentified prod.
3	92	**240**	-	5*10^{-3}	THF	reflux	**241**
4	93	CuTC	-	2*10^{-3}	NMP	0 °C	**242** (48 %) + **241**
5	-	CuTC	-	2*10^{-3}	NMP	r.t.	**242** (88 %)
6	-	CuTC	-	5*10^{-3}	NMP	r.t.	**242** (~20 %)

---

[91] Hodgson, D. M.; Foley, A. M.; Boutlon, L. T.; Lovell, P. J.; Maw, G. N. *J. Chem. Soc., Perkin Trans. I* **1999**, 2911 – 2922.

[92] Brody, M. S.; Finn, M. G. *Tetrahedron Lett.* **1999**, *40*, 415 – 418.

[93] a) Allred, G. D.; Liebeskind, L. S. *J. Am. Chem. Soc.* **1996**, *118*, 2748 – 2749; b) Paterson, I.; Man, J. *Tetrahedron Lett.* **1997**, *38*, 695 – 698.

RESULTS AND DISCUSSION – 3RD GENERATION APPROACH

7	-	PdCl$_2$(PPh$_3$)$_2$	DIBAl-H	1*10^{-3}	DMF	r.t.	No prod/	Pd^0 by DIBAL-H reduction
8	69	PdCl$_2$(CH$_3$CN)$_2$	-	1*10^{-2}	NMP	r.t.	**242** (13 %)	
9	-	PdCl$_2$(CH$_3$CN)$_2$	-	3*10^{-3}	NMP	r.t.	**242** (33 %)	
10	-	PdCl$_2$(CH$_3$CN)$_2$	-	5*10^{-3}	DMF	r.t.	**242** (58 %)	
11	-	PdCl$_2$(CH$_3$CN)$_2$	-	5*10^{-3}	DMF	r.t.	**242** (19 %)	
12	-	PdCl$_2$(CH$_3$CN)$_2$	-	5*10^{-2}	DMF	r.t.	**242** (47 %)	

Table 2: Intramolecular Stille coupling between C-5 and C-6

Even more disappointing was the fact, that all attempts to cyclise **242** in a transannular Diels-Alder reaction had failed. This turned out to be more understandable when later analyses (in particular Diffusion-Ordered Spectroscopy – DOSY) and comparisons with later products assigned **242** to be the dimeric product. (Scheme 65, Figure 24)

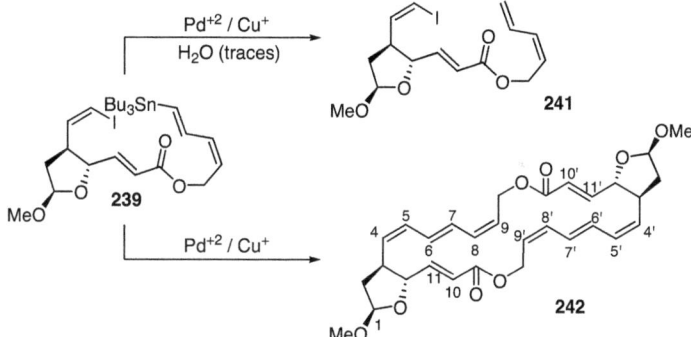

Scheme 65: Products of the intramolecular Stille coupling between C-5 and C-6

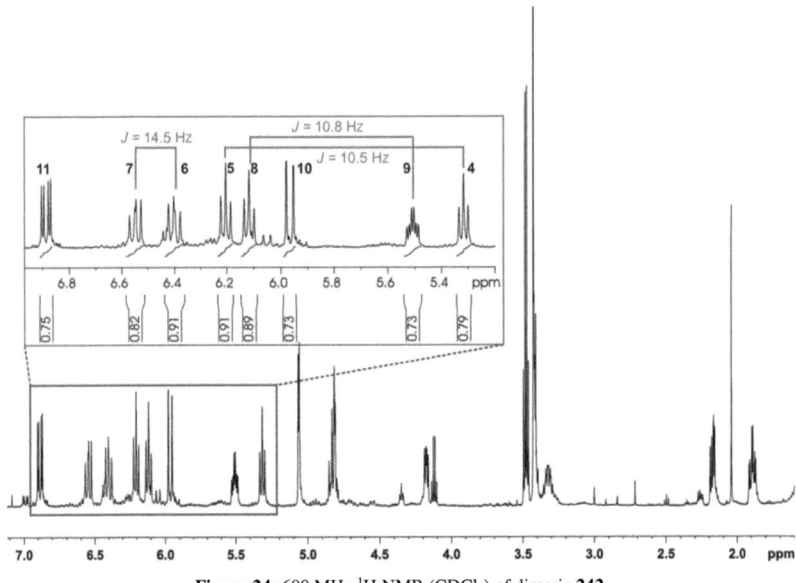

Figure 24: 600 MHz ^{1}H NMR (CDCl$_3$) of dimeric **242**

7.2.2 'Stitching' coupling approach

The previous formation of dimers became more understandable, when a related ring-closing strategy was attempted. Inspired by the so-called 'stitching coupling' in Nicolaou's rapamycin synthesis,[94] we also tried to apply this method to our system. (Scheme 66) In our case, this meant that the *E*-double bond between C-5 and C-8 should be inserted as one individual piece in the form of *E*-distannylethene **243**[95] into a bis-vinyliodide **244**.

---

[94] Nicolaou, K. C.; Chakraborty, T. ; Piscopio, A. D.; Minowa, N, Bertinato, P. *J. Am. Chem. Soc.* **1993**, *115*, 4419 – 4420.

[95] a) Renaldo, A. F.; Labadie, J. W.; Stille J. K. *Org. Synth.* **67**, 86 – 97; b) Bottaro, J. C.; Hanson, R. N.; Seitz, D. E. *J. Org. Chem.* **1981**, *46*, 5221 – 5222.

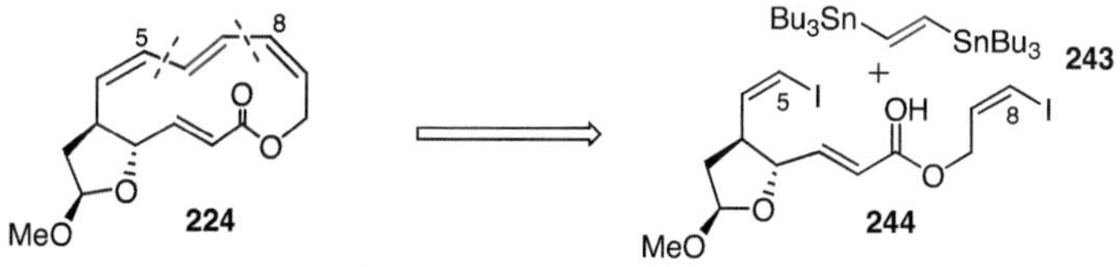

Scheme 66: Stitching approach to the macrolactone

To achieve this, the same core building block **216** could be used again. (Scheme 67) The synthesis of the required Z-iodo-allyl-phosphonate **247** was accomplished as follows: 1,4-Addition of LiI to ethyl propiolate **245** gave exclusively the Z-iodo-ethyl acrylate.[96] DIBAl-H reduction gave the corresponding allylic alcohol **246**, which was then esterified to give phosphonate **247**. IBX-Oxidation of **216a**, followed by Roush-Masamune olefination with phosphonate **247** led to the Z,Z-divinyliodide **244**.

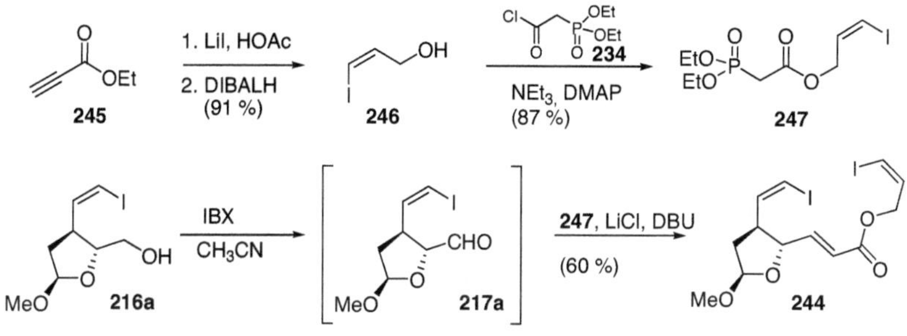

Scheme 67: Synthesis of the bis-vinyliodide – precursor **244** for the stitching coupling

Initial coupling attempts using Pd$_2$dba$_3$/AsPh$_3$ (Table 3, entry 1) did not give any product – on the other hand, on this system, the previously unsuccessful coupling conditions (entry 2) led to the formation of highly UV-active, unknown product **248**. Even higher yields could be obtained when the CuTC-coupling was applied to this system. (entries 3-4).

---

[96] Ma, S.; Lu, X.; Li, Z. *J. Org. Chem.* **1992**, *57*, 709 – 713.

entry	Pd-source	Additives	conc (M)	Solvent	T (°C)	Products/remarks
1	Pd$_2$(dba)$_3$	AsPh$_3$	1*10^{-2}	NMP	45 °C	no product
2	PdCl$_2$(PPh$_3$)$_2$	DIBAl-H	1*10^{-2}	DMF	r.t.	**248** (32 %)
3	CuTC	+1.2 eq. **243**	1*10^{-2}	NMP	r.t.	**248** (54 %)
4	CuTC	+ 2.0 eq. **243**	5*10^{-3}	NMP	r.t.	**248** (38 %)

Table 3: Assessed condition in the 'stitching' coupling

This new product **248** turned out to be quite unstable – upon prolonged standing in CDCl$_3$, a new slightly more apolar spot formed, which seemed to be the result of an isomerisation process. Analysis of the COSY spectra showed that the Z-vinyl group bearing carbon C-5 was still present in **248**; in total, 7 olefinic signals (besides the acetal proton) could be found in the spectrum – this uneven number of double bond signals could only be accounted for with the pseudo-dimer **248**. (Scheme 68) This also explained its high UV-activity and the instability of **248** towards slightly acidic conditions (silica gel, CDCl$_3$), upon which isomerisation was observed.

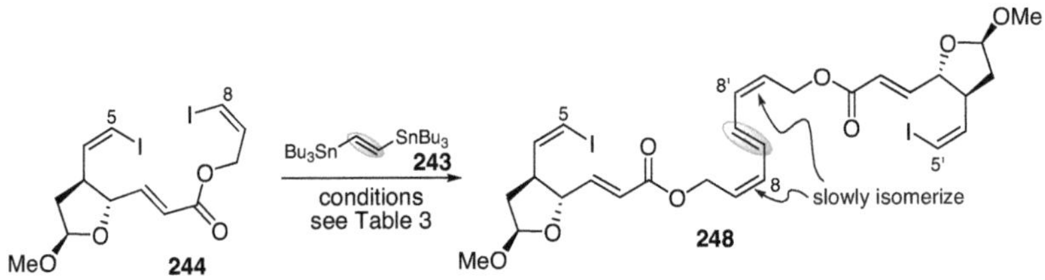

Scheme 68: 'Stitching' coupling leading to pseudo-dimer **248**

The exclusive formation of **248** gives a clear indication of differently reactive sites in the molecule: (Scheme 69) Apparently, the C-8 – iodine bond in **244** is more reactive towards oxidative insertion of the copper(I) – salt. After formation of the vinyl-copper(III) bond, copper species **249** undergoes transmetallation and reductive elimination to give *seco*-compound **239**. A second oxidative insertion, this time at the C-5 – iodine bond, should lead to C-5 – copper(III) – species **250**. However, this oxidative insertion apparently never occurred, as neither a dehalogenated product nor the desired macrocycle were obtained. Instead, metallated **249** rather underwent an intermolecular cross-coupling with *seco*-compound **239** to give the pseudo-dimeric compound **248**.

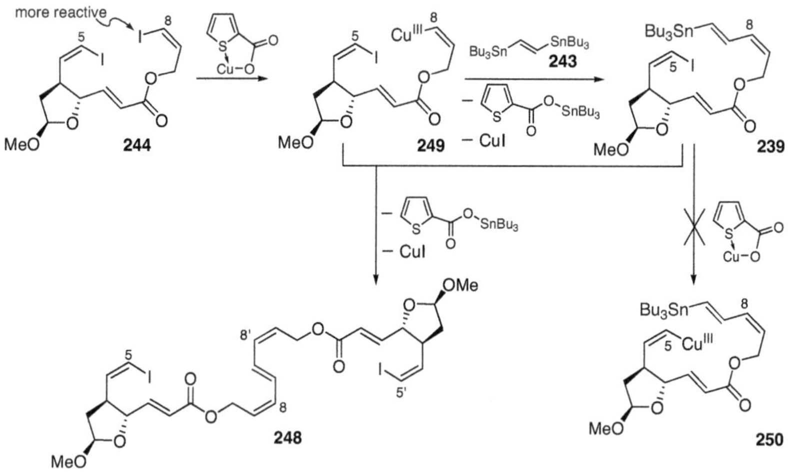

Scheme 69: Kinetic preference of the pseudo-dimer **248**

Interestingly enough, also on this molecule, the vinyl iodides at the positions 5 and 5' proved totally unreactive towards oxidative addition. This could not be altered by adding an excess of the distannane (see Table 3, entry 4). Under the examined reaction conditions, a metal-halogen exchange of the C-5 vinyl iodide never occurred.

From these experiments it could be concluded that

- if (under certain reaction conditions) a metallation at C-5 occurred, the conformation of this species seems to be highly unfavourable for the formation of the 13-membered macrocycle
- the C-8 vinyl iodide is much more reactive towards oxidative insertion than its C-5 counterpart
- an intramolecular coupling should be much easier between C-7 and C-8.

7.2.3 C-7 – C-8 Stille coupling approach

Based on the conclusions drawn from the last experiments, the C-4 – C-7 unit should be assembled first, bearing an appropriate substituent to allow the coupling with the allylic moiety. For such a coupling, the following compounds can be imagined as *seco*-precursors. (Figure 25)

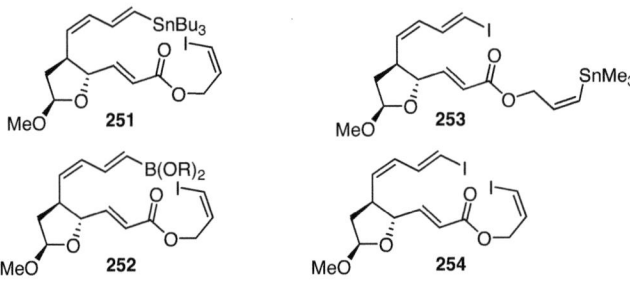

Figure 25: Potential *seco*-compounds

As the Stille-coupling precursor **251** seemed to be most easily available, it was tested first for the coupling. The biggest problem in the synthesis of **251** appeared to be the lability of the vinyl tributyl tin group. Considering the similar reactivity of organotins and organosilicons, we decided to avoid the use of silyl protecting groups when also an organotin group is present in the molecule. Vinyl iodide **216a** was protected as the acetate **255** before a cross-coupling with distannyl ethene **243** was undertaken.[97] After removal of the acetate (now safely possible under reductive conditions), alcohol **256** could be obtained in excellent yield. Oxidation of the primary alcohol followed by Roush-Masamune olefination with phosphonate **247** did not work as well as anticipated, yielding **251** in only 35 % yield besides enyne **261** (see chapter 7.3.2 for a mechanistic discussion). Upon subjection of **251** to previously tested coupling conditions (see Table 2, entry 10), tetraene **224** was obtained in 37%.

---

[97] Stille, J. K. *J. Am. Chem. Soc.* **1986**, *108*, 3033 – 3040.

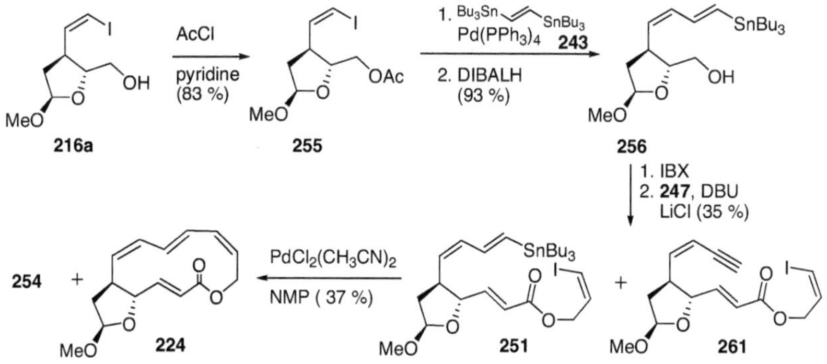

Scheme 70: sp^2-sp^2 Stille coupling between C-7 and C-8 to give **224**

This compound, although contaminated with unidentified minor compound **254**, showed for the first time the allylic H-21 protons as clear AB system with a coupling constant of 15.1 Hz. (Figure 26) The magnetic non-equivalence of these protons indicates them to be fixed in two chemically different environments – which would be the case in a strained macrocycle. The coupling constants of the double bond protons H-4 – H-5, H-6 – H-7 and H-8 – H-9 were determined to be 10.4 Hz, 15.8 Hz and 11.7 Hz, respectively. This is in good correlation with previously observed coupling constants present in acyclic Z,E,Z-trienes. With the two H-21 protons as clear identifiers of product **224**, inspection of NMR spectra of previous coupling attempts showed, that **224** was already formed in an earlier coupling attempt (*cf.* Table 2, entry 1) as the minor product.

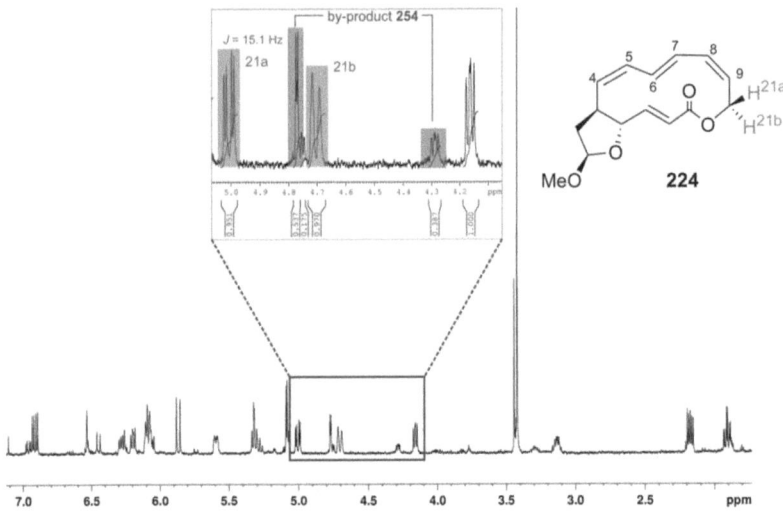

Figure 26: 600 MHz ^{1}H NMR (CDCl$_3$) of 13-membered macrolactone **224**

The structure was confirmed by high-resolution mass spectroscopy, as well as the use of DOSY – NMR. Interestingly, when compared with the diffusion data of dimer **242**, the diffusion rates of **242** fitted better to a trimer – apparently, its stretched structure overproportionally decreased its diffusion speed.

Initial efforts to convert this macrocycle to the Diels-Alder product by heating to 60 °C failed. However, when **224** was heated in toluene-d_8 at 105 °C over one week, a slow conversion could be monitored by NMR. (Scheme 71) This new compound showed, in comparison to the starting material, a surprisingly high polarity on TLC, and did not stain with anis aldehyde, which was so far used as a reliable detecting agent of the lactol group in the molecule. As can be seen in the NMR stack plot, apart from a minor impurity present already in the starting material, **225** was produced as the main compound.

In later optimisations, toluene was replaced by xylenes as solvent (boiling point ~140 °C), which led to a reduction of the reaction time from 168 hours to 36 hours. To avoid oxidation of the double bonds at these elevated temperatures, 2,6 bis-*tert*-butyl-4-methyl phenol (BHT) was added to the reaction mixture. Under these conditions, the TADA product was obtained in 80 % yield.

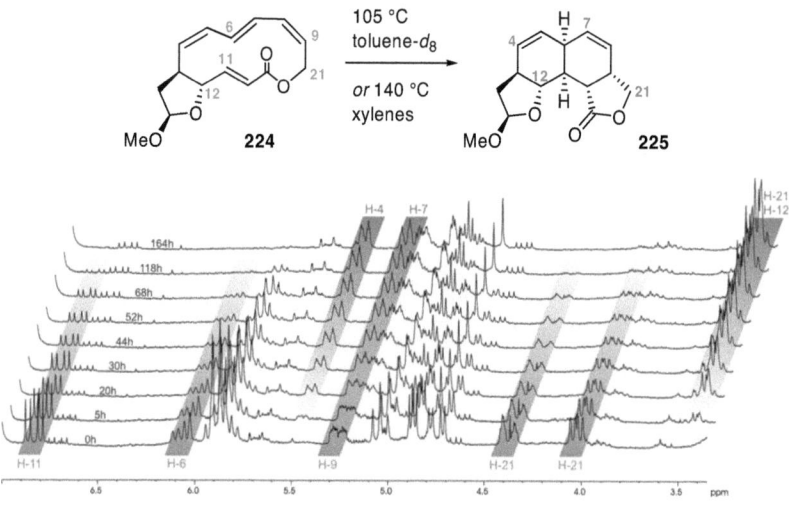

Scheme 71: TADA cyclization of **224**

The ^{1}H NMR of compound **225** shows only 4 olefinic signals, besides the lactol proton at 5.15 ppm. (Figure 27) Furthermore, the signals of the diastereotopic C-21 CH$_2$-group have moved about half a ppm upfield (*cf.* Figure 26), and four additional protons have appeared in the aliphatic region between 2.5 and 3.3 ppm. The proton connectivities could be assigned by DQF-COSY and TOCSY (10 ms mixing time) experiments – the TOCSY experiments were necessary as the coupling constants between H-3 – H-4, H-10 – H-11 and H-9 – H-10 were very small and therefore the crosspeaks were not well resolved in the DQF-COSY spectrum.

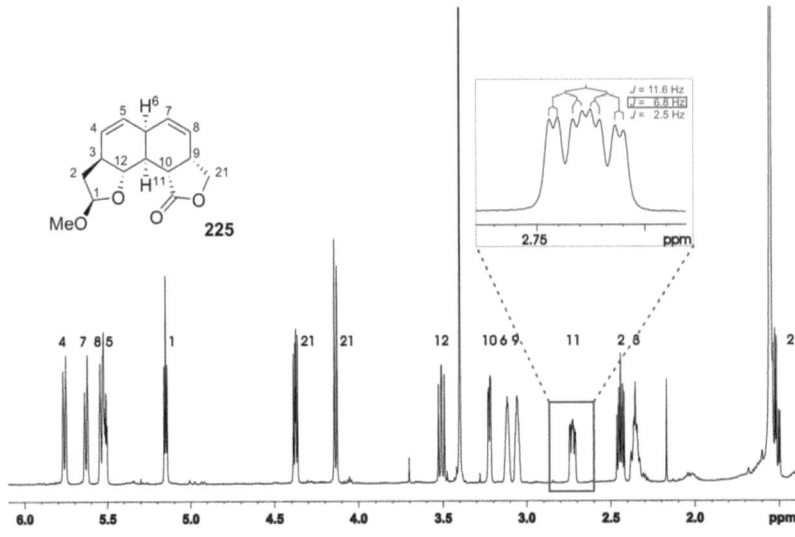

Figure 27: 600 MHz ^{1}H NMR (CDCl$_3$) of TADA product **225**

The stereochemistry of **225** could be assigned by extensive analysis of the NOESY spectra. Key NOE-interactions are shown in Figure 28a – most important are the clear interactions between H-12 and H-9, as well as H-3 and H-11. It should be noted that no direct interaction could be observed with H-6, which would have provided a direct proof of the *cis*-fusion of the hexalin core. Indirect evidence was available from the coupling constants of H-11, where a 6.8 Hz coupling constant could be assigned to the H-11 – H-6 skalar coupling. (Figure 27) However, this coupling constant could not be found on H-6, as this proton only appeared as a unresolved multiplet even at 600 MHz.

Later, one batch of *cis*-hexalin **225** crystallised after column chromatography. Slow crystallisation from hexanes/chloroform yielded crystals suitable for X-ray single crystal diffraction. (Figure 28b) This crystal structure not only proofs the proposed stereochemistry, it also explains the observed NOE-interactions, especially between C-9 and C-12. Finally, the crystal structure is in perfect accordance with the structure previously proposed by molecular modelling (*cf.* Figure 23).

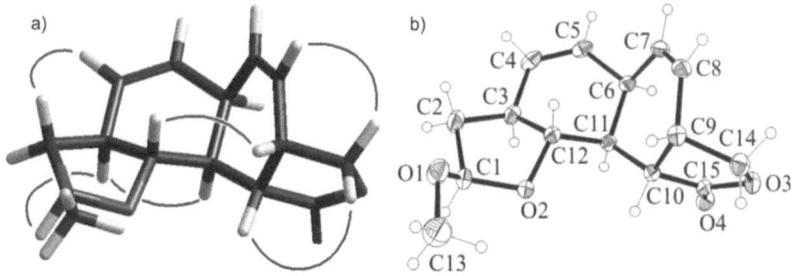

Figure 28: a) Observed NOE interactions on 3D model of **225** b) ORTEP-projection of TADA product **225**

Later experiments, carried out on bigger batches of tetraene, allowed the characterization and identification of a minor, diastereomeric TADA product **226** and the impurity **254**. (Scheme 72) After the Diels-Alder reaction, the TADA products could be easily separated from the impurity, which was now identified as *E,E* diiodide **260**. Although indiscernible on TLC, **260** could not be completely superimposed with the impurity **254** before the thermal Diels-Alder reaction. It is conceivable that the *Z,E*-vinyl iodide was not stable under thermal conditions and isomerised to the *E,E*-isomer. (Scheme 72) Interestingly, the allylic *Z*-vinyl iodide was completely stable under these conditions.

Furthermore, careful chromatographical separation of the TADA products allowed the characterisation of a second diastereomer **226**, present in a ratio of **225**:**226** = 10:1.

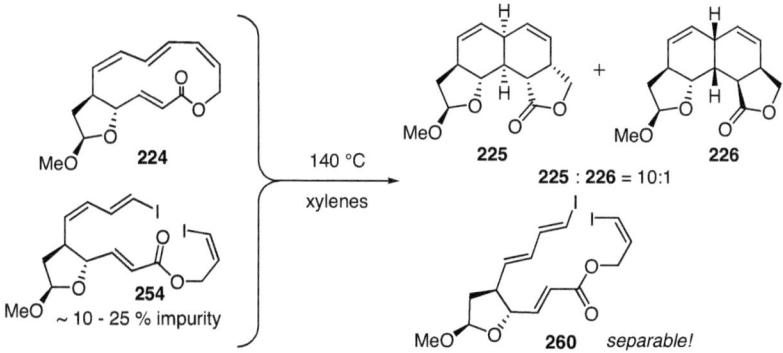

Scheme 72: Detailed investigation of the TADA reaction

The ^{1}H NMR spectrum of **226** was again very well resolved, although some shifts are quite different: (Figure 29) the bridgehead proton H-11 appears further downfield, whereas the other bridgehead proton H-6 has not shifted at all. H-10, on the other side, moved about 0.3 ppm upfield. Both effects are understandable, as these two protons have moved towards (H-10) or away (H-11) from the lactol oxygen.

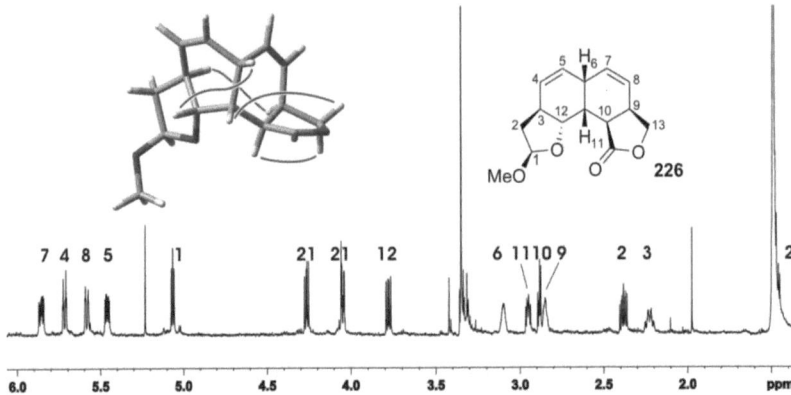

Figure 29: 600 MHz ^{1}H NMR (CDCl$_3$) of TADA by-product **226**, with observed NOE's on 3D-model

From these experiments it can be summarised that a 13-membered macrocycle is indeed accessible by a cross-coupling between C-7 and C-8. The so formed, thermally stable Z,E,Z,E-tetraene macrolactone **224** can undergo a thermal Diels-Alder reaction to give *cis*-hexalin **225** as the main product. However, to use these findings successfully in our total synthesis, two main concerns had to be dispelled before. First, the poor yield of the olefination had to be improved – an idea that seemed quite easily accomplished. Second, the coupling needed improvement – either by using another *seco*-precursor (see Figure 25), or by applying optimised conditions in the coupling itself. Many of the experiments described in the following chapter have been executed in a parallel manner – the order of their listing therefore does not necessarily represent their chronological sequence.

7.2.4 Alternative C-7 – C-8 couplings

Keeping the basic concept of an intramolecular coupling between C-7 and C-8 intact, we decided to evaluate the feasibility and scope of alternative couplings. With this investigation, we hoped a higher yielding coupling reaction could be found.

Furthermore, when using different building blocks, a higher yield in the olefination reaction forming the C-10 – C-11 bond was conceivable.

As a first attempt, we planned to access **252** (Figure 25) where the vinyl tin group is replaced for a boronate, to perform an intramolecular Suzuki-coupling. Inspection of the literature showed that in most cases, the boronates were introduced directly before the coupling, indicating the instable nature of this functional group. However, cyclic, sterically demanding boronate esters have been reported to be stable enough for routine purification methods such as column chromatography.[101]

The installation of the boronate ester was envisioned *via* a terminal enyne: C_2-elongation of *Z*-vinyl iodide **208a** *via* Sonogashira-coupling with TMS-acetylene led to enyne **261** in acceptable yield, followed by basic cleavage of the alkyne-TMS group. The terminally unprotected enyne **262** should then be hydroborated with 9-BBN. Normally, hydroborations of terminal alkynes to the corresponding *E*-alkenes are troublesome reactions, as they always lead to partial formation of the diboryl alkane. This can be avoided, when 2 equivalents of 9-BBN are used deliberately (giving exclusively the diboryl species), followed by quenching of the reaction with benzaldehyde, which reacts selectively with the diborane adduct in a dehydroborylation reaction to regenerate the *E*-vinyl borane **263**.[98] Unfortunately, in our case the application of this method led to decomposition; so the prospected oxidation to boronate **264** with trimethylamine *N*-oxide[99] was not undertaken.

---

[98] Colberg, J. C.; Rane, A.; Vaquer, J.; Soderquist, J. A. *J. Am. Chem. Soc.* **1993**, *115*, 6065 – 6071.

[99] a) Zweifel, G.; Polston, N. L.; Whitney, C. C. *J. Am. Chem. Soc.* **1968**, *90*, 6243 – 6245; b) Ichikawa, J.; Moriya, T.; Sonoda, T.; Kobayashi, H. *Chem. Lett.* **1991**, 961 – 964.

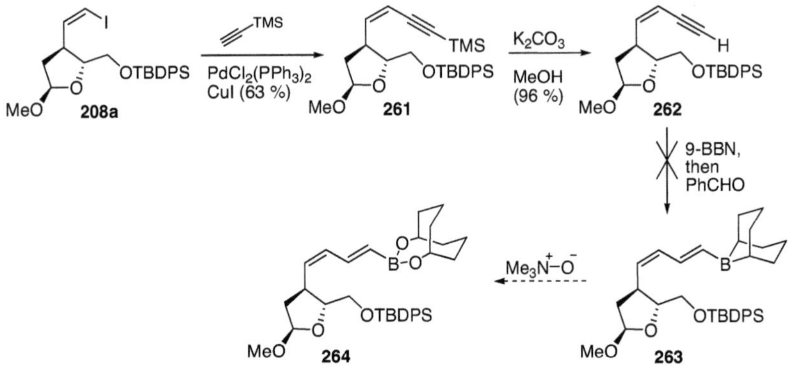

Scheme 73: Attempted synthesis of the *E,Z*-boronate **265**

Alternatively, the desired boronate could be introduced in a cross-coupling using bis-pinacolato-diborane (**268**).[100] The pinacolato-boronate produced in such a manner should also be stable enough for further conversions.

TBDPS-protected *E,Z*-dienylstannane **265** (*vide infra*) was treated with iodine to give *E,Z*-dienyliodide **267**. (Scheme 74) In a first attempt, coupling with diborane **268** gave indeed the boronate ester **269**. However, deprotection of the primary alcohol using TBAF to give **270** was not successful, although such transformations have been reported in the literature.[101]

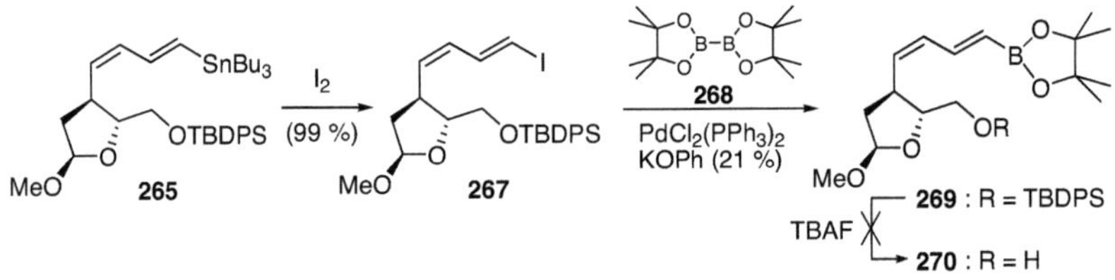

Scheme 74: Cross-coupling boronate introduction

---

[100] Takagi, J.; Takahashi, K.; Ishiyama, T.; Miyaura, N. *J. Am. Chem. Soc.* **2002**, *124*, 8001 – 8006.
[101] Shindo, M.; Sugioka, T.; Umaba, Y.; Shishido, K. *Tetrahedron Lett.* **2004**, *45*, 8863 – 8866.

Given the difficulty of the preparation of the desired boronates at an early stage, we decided to introduce the desired vinyl-metal moiety as late as possible. This could be achieved by installing either a vinyl tin or a vinyl boronate species on the C-8 – C-9 allylic alcohol. In the case of the tin residue, the reaction should be tried with the more reactive trimethyltin group; here, this quite labile group needs not to survive more than one purification by chromatography before the coupling.

Z-Iodo allylic alcohol **246** could be directly metallated and alkylated with trimethyltin choride to yield Z-vinyl stannyl alcohol **271**.[102] (Scheme 75) Because of the highly instable nature of the stannyl alcohol, we did not dare to synthesise the phosphonoacetate **272** *via* the acid chloride (*vide supra*), as residual acid might cleave the C-Sn bond. Instead, we opted for a more neutral esterification method using EDC·HCl as dehydrating agent. The reaction, although quantitative by TLC analysis, yielded after column chromatography (despite addition of 5 % NEt$_3$ to the eluent) only 12 % of the desired vinyl-tin phosphonate **272**, with the destannylated allylic ester as the major by-product.

Synthesis of the analogous Z-vinyl boronate **276** was anticipated to work in analogy. (Scheme 75) To avoid the formation of an unsymmetrically substituted boronate ester, the primary alcohol **246** was protected as TMS-ether **273**. Upon metal-halogen exchange with *t*-BuLi, addition of isopropoxy-pinacolatoborate (**274**) led to formation of a small amount of the desired product, which was unfortunately completely destroyed in a purification attempt. No further attempts towards the synthesis of vinyl borate **276** were undertaken after this failure.

---

[102] Dineen, T. A.; Roush, W. R.; *Org. Lett.* **2004**, *6*, 2043 – 2046.

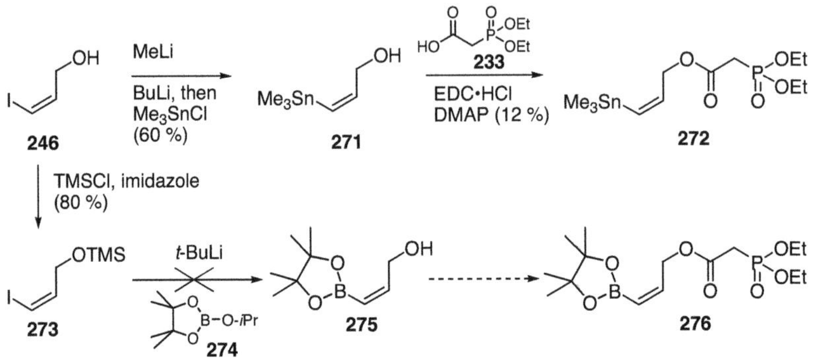

RESULTS AND DISCUSSION – 3RD GENERATION APPROACH

Scheme 75: Synthesis of Z-metallated allylic esters **272** and **276**

Due to the unavailability of **276**, the coupling could only be tried with trimethyl tin phosphonate **272**. To access the desired cross-coupling precursor, *E,Z*-dienyl iodide **267** was deprotected using TBAF to give the free primary alcohol **277**. (Scheme 76) Surprisingly, after deprotection, the corresponding *E,E*-dienyl iodide could also be isolated in variable amounts. The following conversion to the *seco*-compound **253** was achieved in only poor yield, although purification of **253** by column chromatography was this time less problematic in respect to destannylation; isomerisation of the *E,Z*-diene persisted, which partially explains the low yield (37 %). Closure to the macrocycle **224** could be achieved using Pd(PPh$_3$)$_4$/CuCl,[75,102] although in only poor yield.

Besides the conventional coupling methods discussed above, also some "exotic" variations were attempted. In this sense, intermediate aldehyde **278** was olefinated with phosphonate **247** to give divinyldiiodide **254**. The cyclisation attempts by a reductive, Ullman-type coupling using CuTC[103] or in an *in situ* metallation with indium metal, followed by heterogenous Pd-catalysis,[104] gave only isomerised starting material. Besides this dissappointment, comparison of *seco*-compound **254** with the side-product

---

[103] Zhang, S.; Zhang, D.; Liebeskind, L. S. *J. Org. Chem.* **1997**, *62*, 2312 – 2313.
[104] Lee, P. H.; Seomoon, D.; Lee, K. *Org. Lett.* **2005**, *7*, 343 – 345.

obtained in the cross-coupling of **251** (see Scheme 72) comfirmed the proposed structure unequivocally.

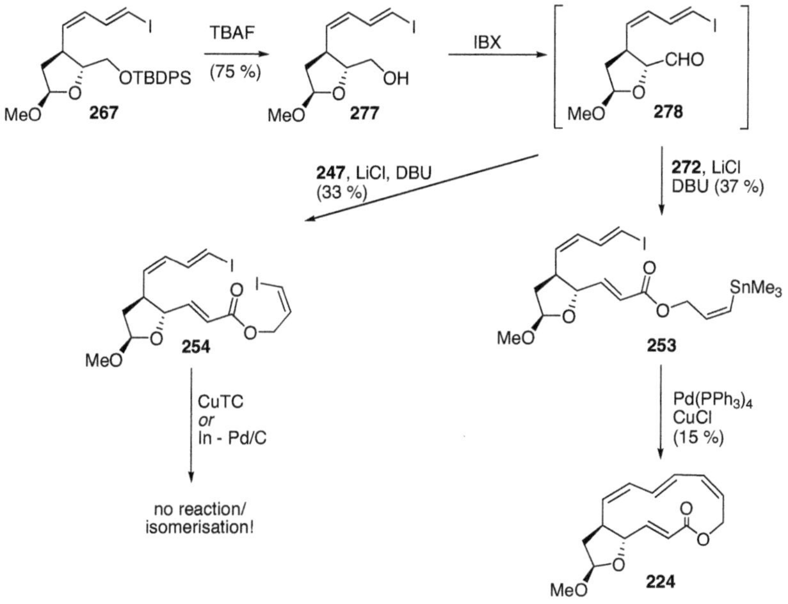

Scheme 76: Alternative cross-coupling attempts

Even though the cyclisation of trimethyl tin bearing *seco*-compound **253** worked, this approach seemed inferior to the one found previously (see chapter 7.2.3) as the configurational lability of the *Z,E*-dienyl iodide made this strategy quite unreliable. Furthermore, cross-coupling with the trimethyl tin residue did not give any improved results.

Another way to facilitate the coupling without the use of the more reactive, but also highly toxic trimethyl stannanes was to expand the ring size of the macrolactone – the larger ring **279** would be beneficial for the coupling of the preceding *seco* compound **280** due to a lower ring strain. (Scheme 77) The disilyl ether/ester **281** would fulfill the requirements for such a tether; after the coupling, this macrolactone can be contracted to original ring size by the treatment with Me$_2$Si(OTf)$_2$.[105] This method was initially developed as a two-step macrolactonisation technique; herein, no purification is undertaken between the formation of the silyl siloxycarboxylate and the Me$_2$Si(OTf)$_2$-induced ring contraction. In our case, the disilyloxy tether should be in place before the ring closure – however there was no indication in the literature about the stability of this disiloxane.

1,2-Bisdimethylsilylbenzene (**282**) was treated sequentially with phosphonoacetic acid **233** at r.t. followed by Z-iodo-allylic alcohol (**246**) at 60 °C, both in the presence of Wilkinson's catalyst. (Scheme 77) The first reaction (using an excess of disilane) worked cleanly (according to NMR), given that the excess disilane was removed in high vacuum. Next, the allylic alcohol **246** was allowed react with the second silane to give **283**. NMR analysis of **283** indicated significant amounts of by-products, so purification was attempted using either strongly deactivated silica gel (5 %) or celite. Despite these precautions, in both cases no product could be isolated – this in turn made it impossible to use **283** in a Roush-Masamune olefination with aldehyde **257** to give the tethered *seco* compound **281**.

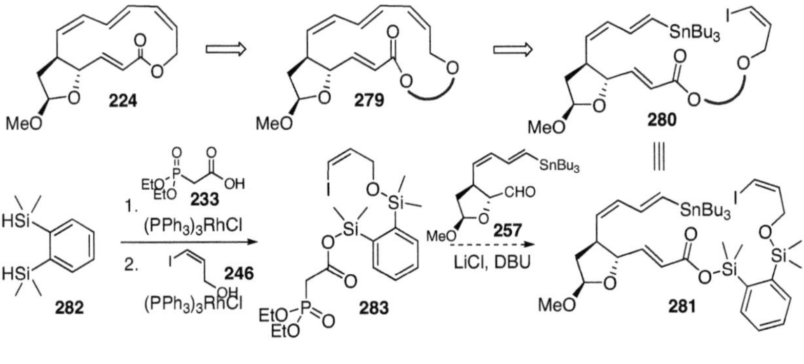

Scheme 77: Attempted use of a phenyldisilane tether

---

[105] Mukaiyama, T.; Izumi, J.; Shiina, I. *Chem. Lett.* **1997**, 187 – 188.

Additionally, also the use of a Negishi-type cross-coupling was investigated. To introduce the even more labile organozinc species necessary for this coupling, enyne **261**, which was obtained as a by-product from sequential IBX-oxidation/Roush-Masamune oxidation of **256** (see chapter 7.3.2), was first treated with the Schwartz reagent to give the vinyl-zirconium species **283**. (Scheme 78) According to literature,[106] this species can be transmetallated with $ZnCl_2$ to give vinyl zinc **285**, which can be cross-coupled under Pd^0-catalysis. Unfortunately, no product formation could be observed under these conditions. It is not clear which transmetallation step did not occur.

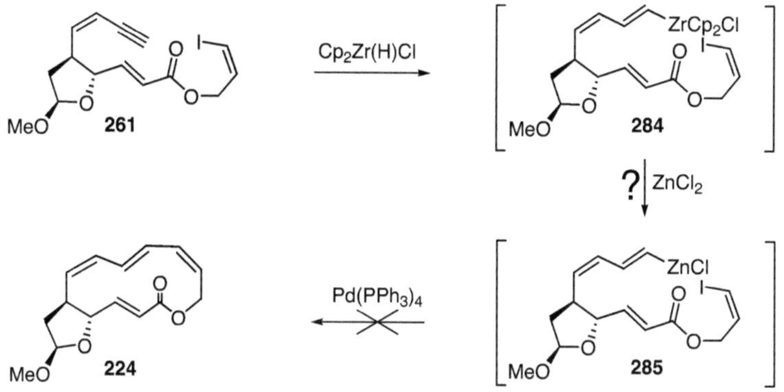

Scheme 78: Attempted Negishi-coupling

All cross-couplings discussed so far form the C-C bond in a C-M to C-Pd transmetallation step, following a reductive elimination. On the other hand, a Heck reaction does not go through such a transmetallation step and would therefore represent a complementary approach. In this system, the Z-vinyl iodide is coupled directly with an unsubstituted Z,E-diene. The most direct method to synthesise this Z,E-diene is *via* a

---

[106] a) for a review on the use of alkenyl zirconocenes, see: Wipf, P.; Kendall, C. *Chem. Eur. J.* **2002**, *8*, 1778 – 1784; b) Panek, J. S.; Hu, T. *J. Org. Chem.* **1997**, *62*, 4912 – 4913; C) Lee, T. W.; Corey, E. J. *J. Am. Chem. Soc.* **2001**, *123*, 1872 – 1877.

Nozaki-Hiyama-Kishi-Peterson olefination:[107] (1-Bromo-allyl)-trimethylsilane (**286**)[108] could be metallated with CrCl$_2$, produced by *in situ* reduction of CrCl$_3$, to give the allyl-chromium(III)-species **287**, which then reacted in high *anti* selectivity with aldehyde **207a** to give the β-hydroxy-silane **288**, which was directly eliminated using potassium hydride. Interestingly, upon elimination, the TBDPS-protecting group was also cleaved to give directly the primary alcohol **289**, although in poor yield.

To try the Heck-coupling, free alcohol **289** was oxidised and olefinated to give the *seco*-vinyl iodide **290**. To our disappointment, under Heck-coupling conditions developed by Jeffery,[109] no product formation could be detected, so this approach was not further investigated.

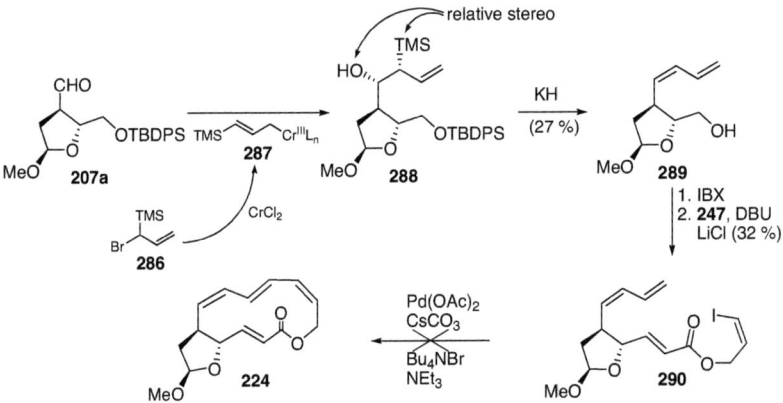

Scheme 79: Formation of *E,Z*-Diene **290** *via* a Nozaki-Hiyama-Kishi-Peterson olefination

---

[107] a) Paterson, I.; Schlapbach, A. *Synlett* **1995**, 498 – 500; b) Paterson, I.; Florence, G. C.; Gerlach, K.; Scott, J. P.; Sereinig, N. *J. Am. Chem. Soc.* **2001**, *123*, 9535 – 9544.

[108] Andringa, H.; Heus-Kloos, Y. A.; Brandsma, L. *J. Organomet. Chem.* **1987**, *336*, C41 – C43.

[109] a) Jeffery, T. *Tetrahedron* **1996**, *52*, 10113-10130; b) Bhatt, U.; Christmann, M.; Quitschalle, M.; Claus, E.; Kalesse, M. *J. Org. Chem.* **2001**, *66*, 1885 – 1893.

7.2.5 Grob-type ring expansion strategy

All cross-coupling approaches discussed so far suffered from either low conversion, instable diene units or problems associated with the preparation and stability of the required vinyl electrophiles. Hence another, different approach was investigated to access the 13-membered tetraene macrocycle. Grob-type fragmentations give access to larger rings in a ring-expansion, concomitant with the selective formation of an *E* or *Z* double bond, dependent on the relative stereochemistry in the starting material. We thought that the application of such methodology[110] would also be possible on our system. Recently, investigations in our group have shown, that also β-hydroxy-δ-lactones can undergo such fragmentations upon treatment with aqueous sodium hydroxide solution.[111]

Inspection of the ring sizes preceding the desired fragmentation indicated that the C-6 – C-7 *E*-double bond in **224** could originate from a fragmentation of the carbonyl – C-7 bond of the intermediate lactol **292**. (Scheme 80) This lactol **292** could be accessed by metallation of the vinyl bromide **293**, followed by intramolecular attack on the δ-lactone. Key intermediate **293** should be accessible in an aldol-type reaction between dihydropyranone **295** and the *Z*-α,β-unsaturated aldehyde **294**. In this approach, no transition-metal catalysed C-C coupling reaction is necessary for the construction of a geometrically defined *Z,E,Z*-triene.

Intrigued by the simplicity of this approach, we decided to test the general feasibility of this idea, as the decisive aldol reaction should occur only with the α-position of the bidentate nucleophile **296** in a regio- and diastereospecific manner on an α,β-unsaturated aldehyde **294** in 1,2 manner. (Scheme 80)

---

[110] Constantieux, T.; Rodriguez, J. *Science of Synthesis,* Vol. 26, **2005** 413 – 462.

[111] Prantz, K. *Diplomarbeit*, Universität Wien, **2005**.

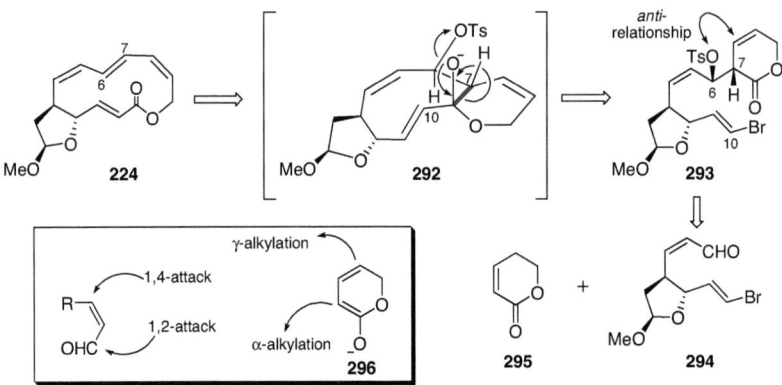

Scheme 80: Fragmentation based approach towards macrolactone **224**

To do this, the reaction between 3,4-dihydropyran-2-one (**295**) and crotonaldehyde (**298**) was used as a test system. (Scheme 81) 3,4-Dihydropyran-2-one (**295**) was synthesised according to a literature procedure[112] from 3-butenoic acid **297** and paraformaldehyde. The fragmentation proposed in Scheme 80 needs an *anti* – relationship between the ring-vinyl group and the allylic hydroxy-function. We thought to accomplish this using boron-enolate chemistry. However, when the boron enolate was formed using (cHex)$_2$BCl and dimethylethylamine, treatment with **298** gave a complex mixture of products.

Another, even milder method to create the enolate, was using a Reformatsky-type reaction.[113] Inspection of the literature showed some metals to be capable of this desired conversions: Cr(II),[114] SmI$_2$,[115] Sn[116] and Zn[117]-induced Reformatsky reactions seemed

---

[112] Nakagawa, M.; Saegusa, J.; Tonozuka, M.; Obi, M.; Ciuchi, M.; Hino, T.; Ban, Y. *Org. Synth.* **56**, 49 – 52.

[113] For reviews on this topic, see: a) Ocampo, R.; Dolbier, W. R. Jr. *Tetrahedron* **2004**, *60*, 9325 – 9374; b) Fürstner, A. *Synthesis* **1989**, 571 – 590.

[114] Wessjohann, L. A.; Scheid, G. *Synthesis*, **1999**, 1 – 36.

[115] a) Obringer, M.; Colobert, F.; Neugnot, B.; Solladié, G. *Org. Lett.* **2003**, *5*, 629 – 632; b) Orsini, F.; Sello, G.; Manzo, A. M.; Lucci, E. M. *Tetrahedron: Asymmetry* **2005**, *16*, 1913 – 1918.

[116] Harada, T.; Mukaiyama, T. *Chem. Lett.* **1982**, 161 – 164.

most promising. Conversion of pyranone **295** to the γ-brominated analogue **299** was accomplished by radical bromination in good yields. Treatment of γ-bromo-lactone **299** with SmI$_2$ at -78 °C followed by addition of **298** led to exclusive formation of the γ-1,2-addition product **301**. On the other hand, the use of either Sn, CrCl$_2$ or Zn gave exclusively the α-1,2 adduct **300** in very low, but also unoptimised yields, the best being the tin(II) enolate (10 %).

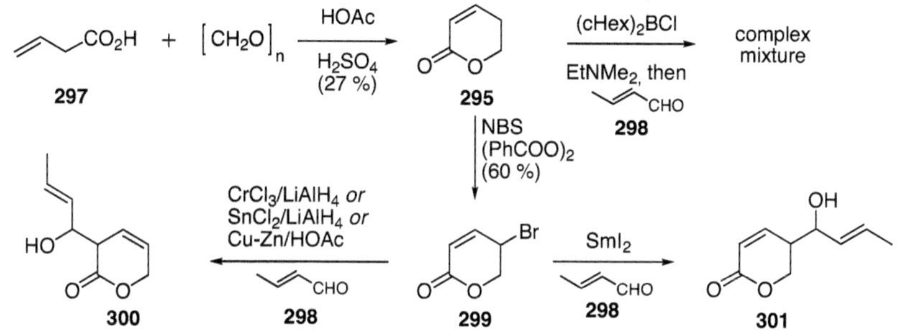

Scheme 81: Evaluation of potential α-1,2-alkylations

Having the general feasibility of this approach in mind, the yield of the aldol reaction was not further optimised, but its diastereoselectivity still had to be determined. (Scheme 82) To do this, lactone **300** was reduced to the triol **302**, which was then treated with anisaldehyde dimethyl acetal **303** under acid catalysis to give the two diastereomeric 1,3-dioxanes **304** and **305**, which could easily be separated by column chromatography and analysed individually. Unfortunately, it turned out that, no matter which metal – CrCl$_2$, Sn or Zn – was used for the Reformatsky reaction, the *anti* : *syn* ratio was constantly 3 : 1, therefore favouring the undesired *anti*-product **304** (this translates to a *syn*-relationsship in **292** – *cf.* Scheme 80)

---

[117] a) Rice, L. E.; Boston, C. M.; Finklea, H. O.; Suder, B. S.; Frazier, J. O.; Hudlicky, T. *J. Org. Chem.* **1984**, *49*, 1845 – 1848; b) Hudlicky, T.; Natchus, M. G.; Kwart, L. D.; Colwell, B. L. *J. Org. Chem.* **1985**, *50*, 4300 – 4306; c) Gurjar, M. K.; Reddy, D. S.; Bhadbhade, M. M.; Gonnade, R. G. *Tetrahedron*, **2004**, *60*, 10269 – 10275.

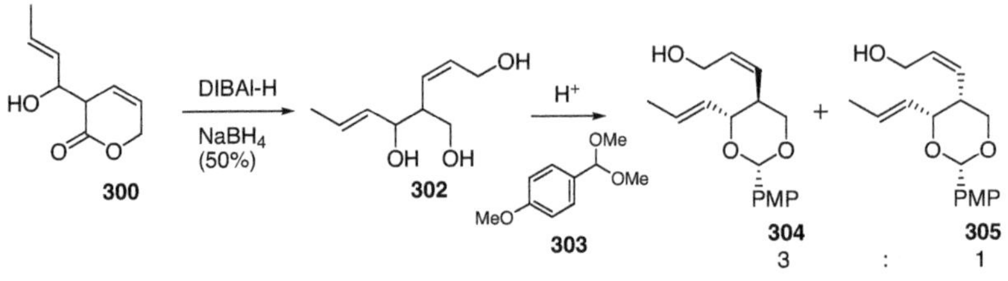

Scheme 82: Assignment of the relative stereochemistry in Reformatsky-aldol product **300**

Parallel to this investigation, also the assembly of the lactol vinyl bromide **294** was started. (Scheme 82) C_2-Elongation of aldehyde **207a** with the Still-Gennari phosphonate **196** gave the Z-α,β unsaturated ester **306** as the sole product. TBDPS-deprotection then gave the primary alcohol **307** in acceptable yields. We thought the C_1-elongation necessary for the formation of the vinyl halogenide was possible in the presence of the C-8 ester group, especially due to the observed instability *i.e.* 'high reactivity', of the intermediate aldehyde. However, after oxidation to the aldehyde **308** using IBX, alkylation using Colvin's reagent[118] gave alkyne **309** only in trace amounts. Therefore, the hydrozirconation/ bromination/reduction to the *E*-vinyl bromide **294** could not be undertaken.

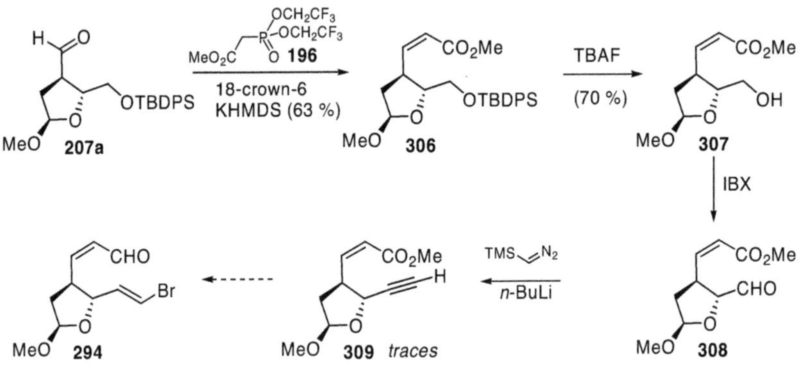

Scheme 83: Assessed synthesis of the C-10 vinyl bromide **294**

---

[118] Colvin, E. W.; Hamill, B. J. *J. Chem. Soc., Chem. Commun.* **1973**, 151 – 152.

Although this approach could be neither disproved nor approved, it had to be put aside, as the encountered problems with this attempt did not allow a fast evaluation of the overall concept.

At this point it was realised, that all alternative methods to construct the macrolactone turned out to be more problematic than the first successful one. Therefore, we focussed to optimise the originally described to macrolactone **224** (see chapter 7.2.3).

7.3   Optimisation of the synthesis of macrolactone 224

For the optimisation of the sequence to the macrocycle, three steps had to be improved:
- Z-selective formation of the C-4 – C-5 double bond: previous experiments have shown the then used Stork olefination (see chapter 6.4.2) to be unreliable, especially upon scale-up.
- C_2-Elongation of the C-9 aldehyde: as can be seen in chapter 7.2.3, the olefination of the dienyl stannyl aldehyde **257** worked in only 35 % yield, compared to 81 % with **217**. Furthermore, the formation of a side-product gave rise to concern.
- Stille-coupling between C-7 and C-8: fine-tuning regarding ligands and additives was needed. The 35 % yield obtained in the first coupling attempt was far from acceptable, so catalysts, additives and solvents had to be screened.

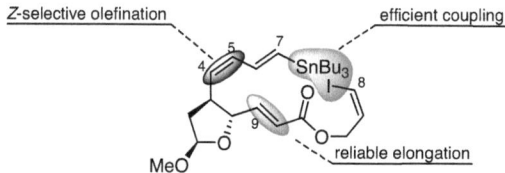

Figure 30: Potential optimisation targets

7.3.1 Alternative Synthesis of the Z-C-4 – C-5 double bond

Due to the problems associated with the preparation of the vinyl iodide **208** and the large amounts of *E*-1,2-bis-(tributylstannyl)-ethene **243** required in the synthesis of the *Z,E*-dienylstannane **256**, we were looking for alternative syntheses. Luckily, such a methodology was disclosed by Brückner *et al.*[119] at that same time, who reported that the (Sylvestre) Julia olefination of vinyl-tin bearing sulfone **313** reacted with a variety of aldehydes in much higher Z-selectivity than usually observed with such sulfones.[120]

The preparation of sulfone **313** was accomplished in a 2 step sequence starting from previously used *E*-3-(tributylstannyl)-allylic alcohol **188**.[68] (Scheme 84) Mitsunobu thioetherification with 2-mercaptobenzothiazol **311** gave the corresponding sulfide **312** in good yields. Oxidation of **312** to **313** was possible in an ammonium molybdate catalysed H_2O_2-oxidation. To assure the reproducibility of the following olefination, a purification method of the brown, oily residue had to be found. Literature reported[119] **313** to destannylate upon column chromatography – as an alternative, we found purification by recrystallization to be possible in methanol, giving sulfone **313** as waxy, yellow crystals.

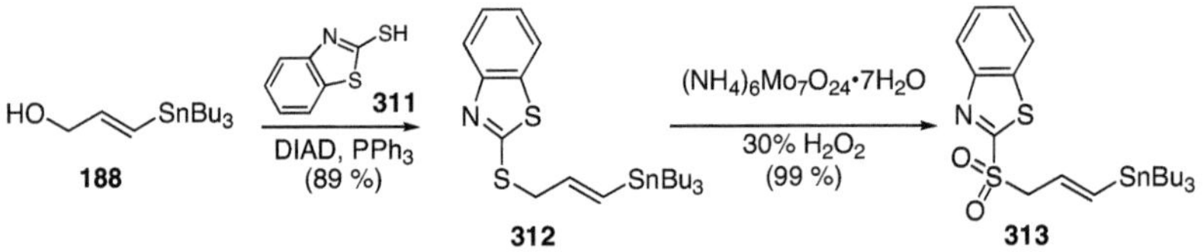

Scheme 84: Synthesis of allylic sulfone **313**

The first olefination attempt with aldehyde **207a** and sulfone **313** under Barbier-conditions using KHMDS in THF gave the desired dienylstannane **265/266** in a *Z,E* : *E,E* ratio of 2 : 1. (Scheme 85)

---

[119] Sorg, A.; Brückner, R. *Synlett* **2005**, 289 – 293.
[120] Blakemore, P. R. *J. Chem. Soc., Perkin Trans. 1* **2002**, 2563 – 2585.

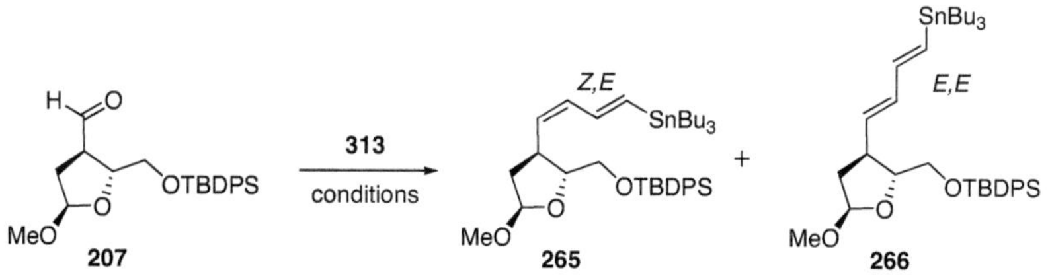

Scheme 85: Z-Selective Julia-olefination

The subsequent optimization regarding excess of sulfone, excess of base, speed of addition and reaction temperature is summarised in Table 4. Following the initial findings by Brückner, the use of toluene as solvent (entry 2), led to a higher $Z,E : E,E$ selectivity, albeit with reduced yield when compared to the same reaction in THF (entry 1). When this reaction was conducted under premetallation conditions (entry 3), a slightly improved, but not reproducible yield was seen – apparently, the controlled formation of the metallated sulfone is the key to success in the optimisation of this reaction. Using an excess of base, in relation to the sulfone (entries 4, 5) proofed undesirable, as can be seen by the diminished yield, especially in entry 5. Stepping back to the higher yielding reactions in ethereal solvents, different methods to improve the selectivity were undertaken: Addition of 18-crown-6 (entry 6) only affected the yield negatively, while a large excess of sulfone and base led to 76 % yield with an unchanged Z/E-selectivity (entry 7). Switching the solvent to DME (entry 8) turned out to further alter the Z/E-selectivity in the unwanted direction. Having excluded the option of solvent and base variation, we started to optimise the addition rate of the base (entries 9-13). Adding KHMDS over 2 hours *via* syringe pump (entry 9) led to a minute improvement, whereas addition over 3 hours led to a decrease of yield. However, when operating at a lower concentration, the optimal addition time was found to be 1 hour, where the yield obtained was 61 %. Besides a constant addition, also long reaction times at -78 °C were necessary to obtain good results. A large excess of sulfone and base is not necessary to obtain good yields.

Entry	eq. base	eq. sulfone	c [M]	Addition time	Solvent	Yield	Z:E : E:E
1	1.12	1.12	0.15	5 min	THF	57 %	2 : 1
2	1.12	1.12	0.15	5 min	toluene	42 %	9 : 1
3	1.6	1.6	0.02	5 min[a]	toluene	52 %[b]	9 : 1
4	1.8	1.6	0.1	5 min	toluene	31 %	9 : 1
5	2.0	1.6	0.1	2 min	toluene	20 %	2 : 1
6	1.1	1.1[c]	0.1	5 min	THF	48 %	2 : 1
7	2.85	2.85	0.1	5 min	THF	76 %	2 : 1
8	1.45	1.45	0.1	1 hour[d]	DME	74 %	1 : 2
9	1.6	1.6	0.15	2 hours[d]	toluene	45 %	9 : 1
10	1.6	1.6	0.05	3 hours[d]	toluene	34 %	9 : 1
11	1.35	1.35	0.05	30 min	toluene	52 %	9 : 1
12	1.45	1.45	0.05	2 hours[d]	toluene	55 %	9 : 1
13	**1.17**	**1.17**	**0.05**	**1 hours[d]**	**toluene**	**61 %**	**9 : 1**

a) sequential addition of KHMDS, followed by the aldehyde; b) not reproducible
c) addition of 1eq. 18-crown-6 to the reaction mixture d) added by syringe pump

Table 4: Optimisation of the Julia-Olefination

This surprisingly high Z-selectivity in the (Sylvestre)-Julia olefination is again due to the presence of the vinyl-tin group in the molecule. Normally, the E/Z-selectivity in the reaction of alkyl BT-sulfones with alkyl aldehydes is governed solely by the nucleophilic addition of the sulfone to the aldehyde. After formation of the β-hydroxy-sulfone, the oxygen anion attacks on the heterocyclic imine functionality, which results in a Smiles-rearrangement[121] and leads to the β-hydroxy-sulfenate. This then undergoes an E_2-elimination leading to the double bond, its E/Z-ratio depending solely on the *erythro/threo* ratio of the initial addition. When β,γ-unsaturated sulfones are used, this situation is changed, as the addition process is now reversible, leading to an increased formation of the Z-olefin. This is due to the lower activation barrier for the Smiles-rearrangement of *threo*-adduct **315** to **316** (leading to the Z-olefin), as the *erythro* adduct **317** suffers from a 1,2 strain. This situation seems to be augmented when a

---

[121] Truce, W. E.; Kreider, E. M.; Brand, W. W. *Org. React.* **1970**, *18*, 99.

vinyl-tin compound is present on the sulfone. Either the repulsive gauche-interaction in **317** is even stronger, or the *threo*-adduct is preferred due to electronic properties of the allyl-tin moiety.

In correlation to the literature data, the formation of the *E*-isomer is augmented with increasing polarity of the solvent (see Table 4, entries 2,7,8). It is unclear however, it this is due to a higher preference of the *erythro*-adduct, or a change in the elimination mechanism of the Smiles-products.

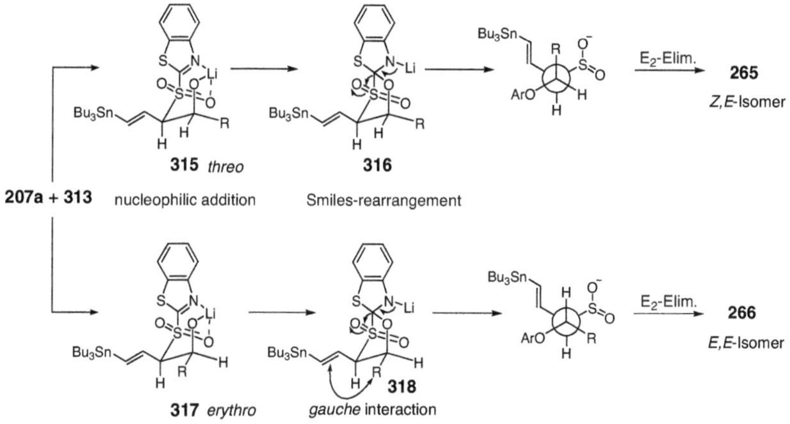

Scheme 86: Mechanism of the modified Julia-olefination

7.3.2 Reliable C-11 – C-10 Olefination

In the course of the preceding optimisation procedure it was realised that the assumed instable vinyl stannane was quite sturdy, at least to basic conditions. Therefore, deprotection of the primary TBDPS ether **265** was tried using TBAF, which led to free alcohol **256** in good yields, with only little amounts of destannylated diene **289**. (Scheme 87) Apparently, the rate of fluorine-induced destannylation is magnitudes slower than the corresponding O-desilylation. The next step that needed improvement was the sequential IBX oxidation/ Roush-Masamune olefination. In the first attempt to access the *seco*-stannyl iodide **251**, it was obtained in 35 % yield, giving alkyne **261** as surprising by-product. Apparently, besides the oxidation to the carbonyl, another reaction had happened which resulted in the elimination of the tributyltin species. In both cases the reaction was terminated by olefination with phosphonate **247**. As we first suspected the elevated temperature in the IBX-oxidation to be responsible for this second elimination reaction, we thought to use the homogenous Dess-Martin oxidation at r.t. instead. However, after precipitation of the periodinane with Et$_2$O, the crude mixture was examined and determined to be pure aldehyde **259**.

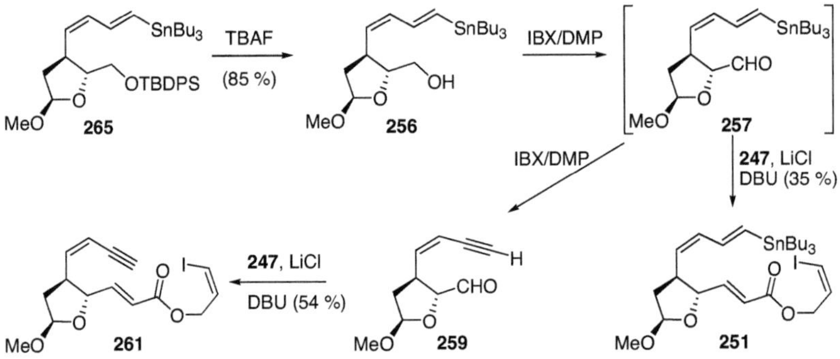

Scheme 87: Hypervalent-iodine mediated formation of side-product

A possible explanation of this strange side-reaction could be found upon inspection of the literature. Nicolaou et al.[122] reported that under specific conditions, ketones could be oxidised to the corresponding α,β-unsaturated carbonyls. They proposed an SET mechanism for this transformation, as outlined in Scheme 88. Based on this mechanism, after oxidation to the ketone **319**, the tautomeric enol **319A** can add again to IBX and form a new complex **319B**. In this complex, a single-electron-transfer occurs, resulting in a radical in the α-position of the carbonyl (**319C**), which can tautomerise to a structure with a cation α to the oxygen (**319D**). Upon E$_1$-elimination of H$^+$, the α,β-unsaturated ketone **320** is formed, along with the reduced iodosobenzoic acid.

---

[122] a) Nicolaou, K. C.; Baran, P. S.; Zhong, Y.-L.; Barluenga, S.; Hunt, K. W.; Kranich, R.; Vega, J. A. *J. Am. Chem. Soc.* **2002**, *124*, 2233 – 2244; b) Nicolaou, K. C.; Montagnon, T.; Baran, P. S.; Zhong, Y.-L. *J. Am. Chem. Soc.* **2002**, *124*, 2245 – 2258.

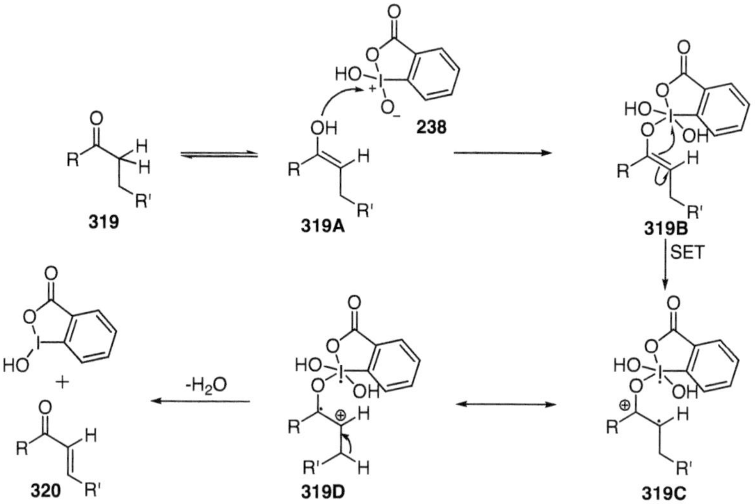

Scheme 88: SET-mechanism proposed by Nicolaou[122]

This proposed mechanism was supported by the observation that cyclopropane aldehyde **322**, upon treatment with IBX, was oxidised to the double unsaturated aldehyde **323**. (Scheme 89) This is explained by the high tendency of the intermediate cyclopropyl radical **322A** to fragment in a homolytic manner. Further support for this mechanism was given by interpretation of kinetic data, which were in favour of the SET mechanism.

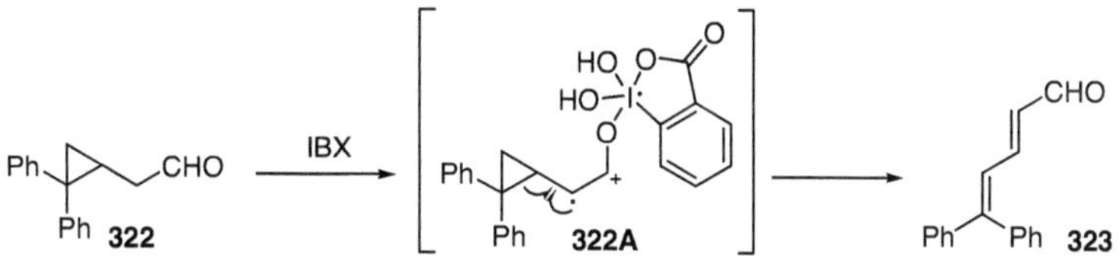

Scheme 89: Evidence of a radical intermediate

Transferring this mechanism to our system results in the following picture: First, the primary alcohol gets oxidised to the aldehyde **257**, which, after tautomerization to **257A**, forms a second complex with IBX. (Scheme 90) After an SET reaction, the initially formed biradical **257B** reacts in a 1,5 H-shift – the so formed vinyl-radical then stabilises itself by expulsion of a tributyltin radical and forms the enyne aldehyde **259**.

This mechanism seems quite likely, as radical 1,5-H shifts are often encountered, and the tributyltin radical is a very good leaving group. The serendipitous observation of this fragmentation is another strong argument in favour of the proposed radical mechanism. With this system, the existance of a radical intermediate could be directly proved if a ^{119}Sn CDNIP effect was observed.[123] Regrettably, such experiments could not be carried out during this Ph.D thesis.

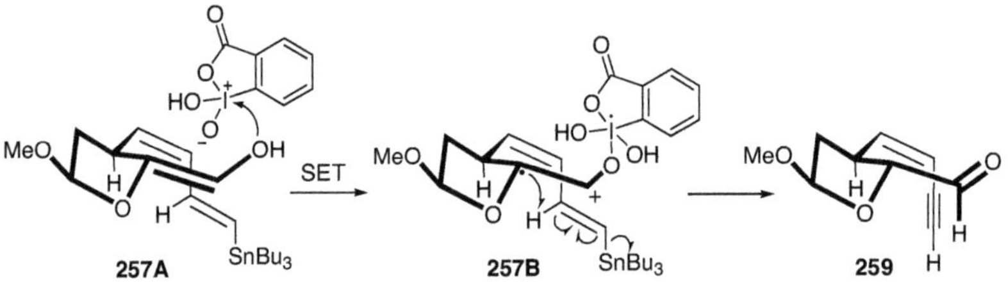

Scheme 90: Proposed formation of the enyne **259**

To avoid this side-reaction, the oxidation of alcohol **256** to aldehyde **257** had to be accomplished in a different manner. Aware of the high lability of **257** we decided to implement a tandem MnO_2-olefination reaction.[81] Although saturated alcohols are normally unreactive towards oxidations with MnO_2, a minute amount of aldehyde seems to be present in the equilibrium, which gets shifted by *in situ* trapping of the aldehyde with a stabilised Wittig-reagent. Application of this tandem reaction using ethyl acetate triphenylphosphoranylen **138** led to formation of the ethyl ester **324** in 40 % yield. (Scheme 91) To convert this product further, the ester **324** had to be saponified to the acid **325**. Not unexpectedly, the free acid in the molecule was not compatible with the vinyl-tin group, resulting in a multitude of signals in the olefinic region of the ^{1}H NMR. Therefore, the same reaction was tried with an appropriately functionalised stabilised Wittig-reagent. This was prepared by condensation of Z-iodo-allylic alcohol **246** with bromoacetic acid (**326**), followed by S_N2 reaction with PPh_3. The required Wittig-reagent was formed by deprotonation of the phosponium salt **328** with NaH,

---

[123] Kruppa, A. I.; Taraban, M. B.; Shokhirev, N. V.; Svarovsky, S. A.; Leshina, T. V. *Chem. Phys. Lett.* **1996**, *285*, 316-322.

either *in situ* or immediately before the experiment. Unfortunately, the previously successful tandem oxidation/olefination sequence did not yield the desired product **251**.

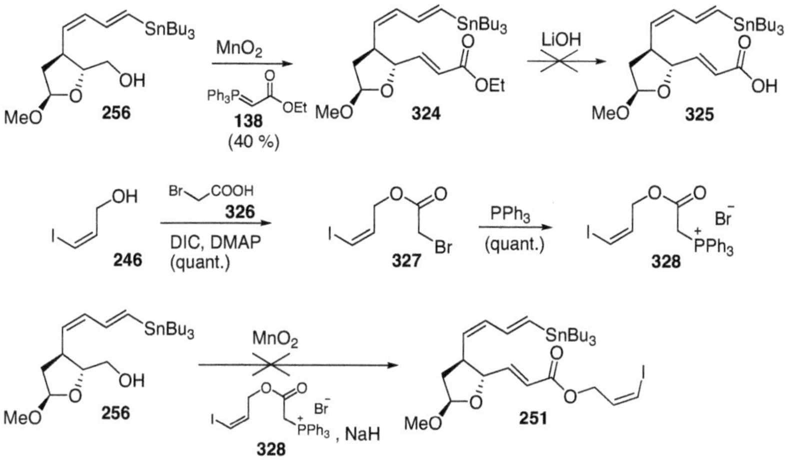

Scheme 91: Attempted tandem oxidation-olefination sequence

Out of desperation, a different idea was developed: Due to the high instability of the aldehyde, we thought to circumvent this stumbling block by first creating the C-C bond in an alkylation step, and form the double bond of the dienophile later, possibly right before the TADA reaction in a one-pot sequence. Therefore, we converted the primary alcohol **256** to tosylate **329**, which should then be alkylated with (phenylsulfinyl)-acetic acid ethyl ester (**330**). (Scheme 92) However, upon subjection to alkylation conditions, besides starting material and a small amount of alkyl iodide, formation of **331** was not observed.

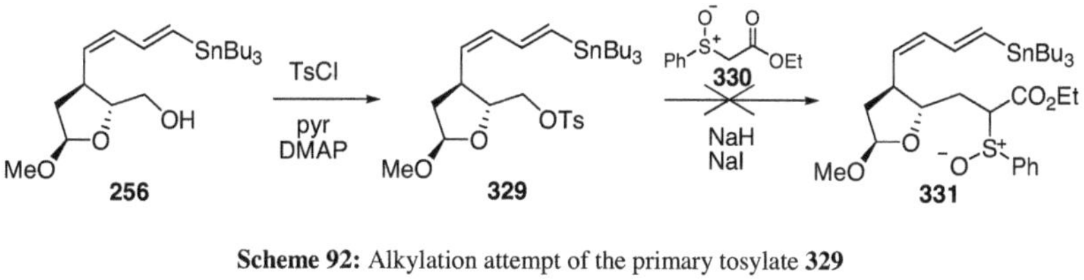

Scheme 92: Alkylation attempt of the primary tosylate **329**

As most of the remaining direct oxidation/olefination sequences seemed too harsh for aldehyde **257**, as a last resort we tried to combine a TPAP-oxidation with the Roush-Masamune olefination. Such a combination is unprecented, although the successful application of a sequential TPAP-oxidation/unstabilised Wittig olefination is known in the literature.[124] To our delight, the oxidation of alcohol **256** in DCM with tetra-n-propylammonium perruthenate (TPAP) using N-methyl-morpholine-N-oxide (NMO) could be followed conveniently by TLC. (Scheme 93) After filtration through a plug of silica and solvent exchange for acetonitrile, the Roush-Masamune olefination worked in the usual reliable manner to yield the desired E-α,β-unsaturated ester **251** in 75 %. Optimisation attempts showed that the overall yield of this reaction was mostly dependant on the speed at which intermediate aldehyde **257** was produced – no reaction occurred at 0 °C, and the reaction was decelerated when water was present in this transformation. Addition of freshly dried molecular sieves to the reaction mixture normally guaranteed a fast conversion.

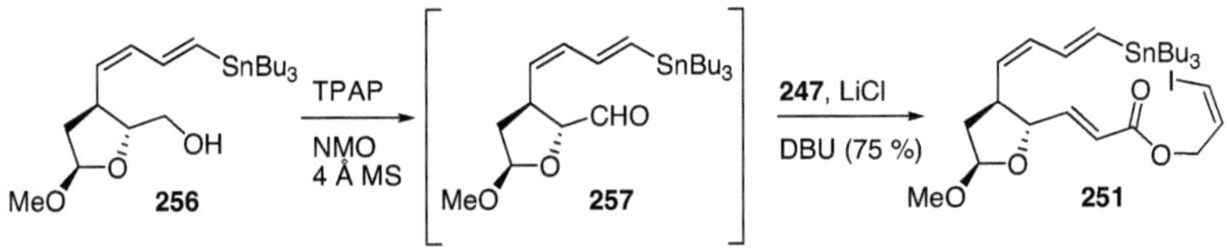

Scheme 93: TPAP-oxidation – olefination sequence

Having now solved both problems regarding the selective and efficient construction of the C-4 – C-5 and C-11 – C-10 double bonds in a geometrically pure manner, the last remaining hurdle was the intramolecular Stille-coupling between C-7 and C-8. It should be noted that this step had now turned from being the first and most obvious choice (from the point of conditions tried) to be the final option (considering that all possible alternatives had already failed). In this conscience, this last problem was tackled.

---

[124] MacCoss, R. N.; Balskus, E. P.; Ley, S. V. *Tetrahedron Lett.* **2003**, *44*, 7779 – 7781.

7.3.3 Optimisation of the C-7 – C-8 intramolecular Stille-coupling

In this intramolecular Stille reaction, the big loss of molar mass (679 Da to 262 Da) necessitates an efficient coupling reaction for the accumulation of useful amounts of material. Having excluded all possible alternatives, improvement of the coupling needed to be done systematically. (Scheme 94, Table 5) We started the optimisation of the Stille coupling of **251** from the previously found conditions. (*cf.* Scheme 70) First it was shown that the yield of the macrocyclic monomer **224** was strongly dependent of the concentration, and therefore high-dilution conditions were applied throughout all couplings. As the impurity **254** was initially attributed to be a mono-chlorinated derivative, the coupling was executed with a chlorine free Pd-source (entry 3), although without any improvement regarding yield or absence of by-product **254**. Coupling with CuTC gave the desired product only in small quantities, along with other undefined by-products (entry 4). When $Pd_2dba_3 \cdot CHCl_3$ (entry 5) was used as a Pd-source, only the desired product was formed, but this time the dba ligand contaminated the product. The use of the bidentate dppf ligand seemed to considerably slow down the conversion, resulting in a low yield (entry 6). A clean product was produced when $Pd(PPh_3)_4/CuCl$ (entry 7) was used, albeit again in low yield. The fluoride-accelerated conditions reported by Baldwin[125] turned out to be ineffective in our case (entries 8, 9). Interestingly, the conditions designed for the coupling of vinyl iodides (entry 8) gave exclusively the dimeric product! When another highly effective system, $Pd(o\text{-furyl})_3/CuI$ was used, a promising yield of 39 % could be obtained (entry 10), although reproducibility was not fully granted. As the influence of CuI in such couplings is reported to be relatively small,[126] both $P(o\text{-furyl})_3$ and $AsPh_3$-ligands were tested without the CuI addition (entries 11, 12) and turned out to be equally effective. When the solvent was changed to less polar THF, using $Pd(PPh_3)_4$ with LiCl as equimolar additive, a very fast and quantitative conversion was observed, although no product could be isolated at all! (entry 13) This situation could be remedied when *n*-

---

[125] Mee, S. P. H.; Lee, V.; Baldwin, J. E. *Angew. Chem. Int. Ed.* **2004**, *43*, 1132 – 1136.
[126] Farina, V.; Krishnan, B. *J. Am. Chem. Soc.* **1991**, *113*, 9585 – 9595.

butylammonium diphenylphosphinate[127] was added to the reaction mixture as a tin-scavenger, which allowed the isolation of tetraene **224** in 61 % yield (entry 14).

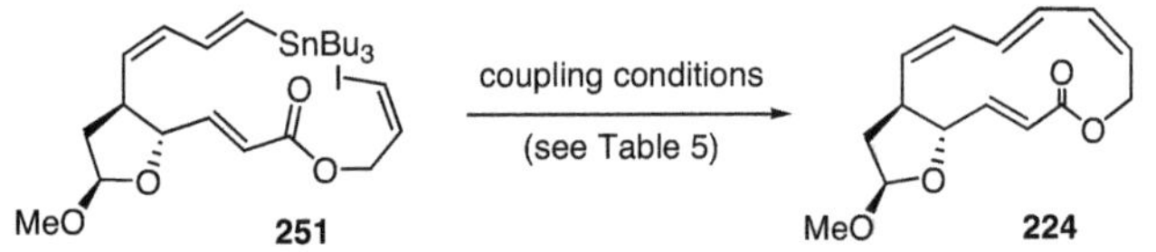

Scheme 94: C-7 - C-8 Stille coupling

entry	Ref.	Pd-source	Additives	c [M]	Solvent	Temp	Yield
1	69	PdCl$_2$(CH$_3$CN)$_2$	-	4*10^{-3}	NMP	r.t.	37 %^{a}
2	-	PdCl$_2$(CH$_3$CN)$_2$	-	2*10^{-3}	NMP	r.t.	25 %^{a}
3	-	Pd(PPh$_3$)$_2$(OAc)$_2$	-	1*10^{-3}	NMP	r.t.	5 %^{a}
4	93	-	CuTC	1*10^{-3}	NMP	r.t.	~10 %
5	91	Pd$_2$dba$_3$·CHCl$_3$	AsPh$_3$	1*10^{-3}	NMP	r.t.	15 %^{b}
6	-	PdCl$_2$dppf	-	1*10^{-3}	NMP	r.t.	17 %^{a}
7	75	Pd(PPh$_3$)$_4$	CuCl	1*10^{-3}	THF/DMSO	r.t.	25 %
8	125	Pd(PPh$_3$)$_4$	CuI, CsF	4*10^{-3}	DMF	r.t.	33 %^{c}
9	125	PdCl$_2$	P(t-Bu)$_3$, CuI, CsF	1*10^{-3}	NMP	60 °C	19 %
10	76	Pd$_2$dba$_3$·CHCl$_3$	CuI, P(o-furyl)$_3$	1*10^{-3}	NMP	r.t.	39 %^{b}
11	126	Pd$_2$dba$_3$·CHCl$_3$	P(o-furyl)$_3$	1*10^{-3}	DMF	60 °C	46 %^{b}
12	126	Pd$_2$dba$_3$·CHCl$_3$	AsPh$_3$	1*10^{-3}	DMF	60 °C	46 %^{b}
13	128	Pd(PPh$_3$)$_4$	LiCl	1*10^{-3}	THF	40 °C	0 %^{d}
14	129	Pd(PPh$_3$)$_4$	LiCl, n-Bu$_4$NPh$_2$PO$_2$	1*10^{-3}	THF	40 °C	63 %

a) Divinyliodide **254** produced as by-product b) product impurified by dba-ligand
c) exclusive formation of dimer **242** d) full conversion observed on TLC

Table 5: Optimisation of C-7 - C-8 Stille coupling

---

[127] Srogl, J.; Allred, G. D.; Liebeskind, L. S. *J. Am. Chem. Soc.* **1997**, *119*, 12376 – 12377.
[128] Garg, N. K.; Hiebert, S.; Overman, L. E. *Angew. Chem. Int. Ed.* **2006**, *45*, 2912 – 2915.
[129] Vaz, B.; Dominguez, M.; Alvarez, R.; de Lera, A. R. *J. Org. Chem.* **2006**, *71*, 5914 – 5920.

After this extensive optimisation procedure, the following picture can be drawn from the examined sp²-sp²-coupling. (Scheme 95) The oxidative insertion of the Pd⁰-catalyst into the C-I bond to form the vinyl-Pd^{II}-species **333** (step 1) seems to be unproblematic in all cases. However, the transmetallation step (step 2) to form the bis-alkenyl-Pd^{II} species **335** appeared to be the problematic step, either for entropic (*i.e.* the accessible conformations of the diene chain did not allow such reaction) or enthalpic reasons (*i.e.* ring strain makes the formation palladacycle **335** unfavourable). The final reductive elimination seemed unproblematic again to yield the macrolactone **224** and the recycled Pd⁰-species.

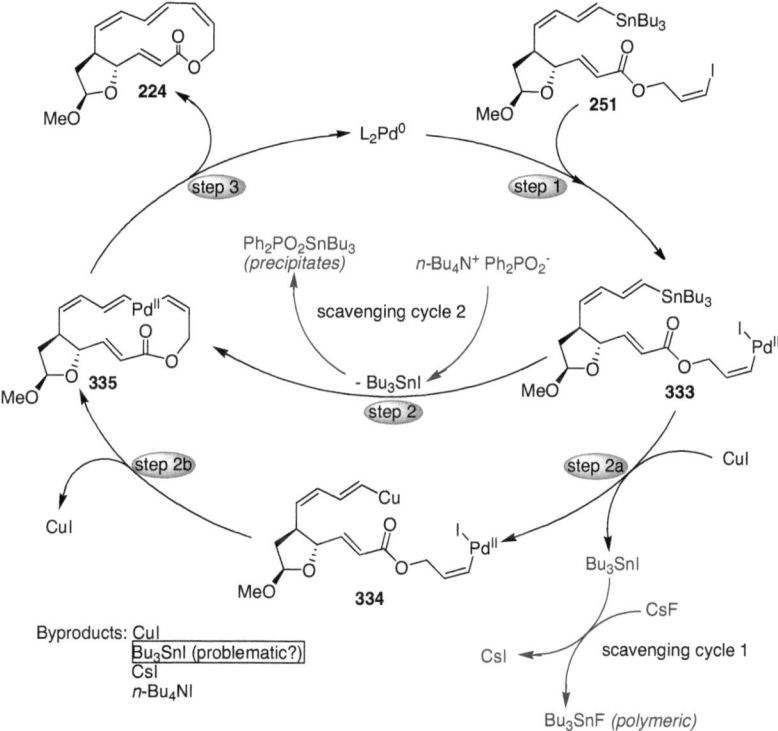

Scheme 95: Mechanistic outline of the intramolecular Stille-Coupling

Based on this rational, copper salts can be added to the reaction mixture. Addition of copper salts to the reaction mixture (in a polar medium) leads to a transmetallation of the vinyl tributyl tin species to a vinyl copper species (step 2a). (Scheme 95) Copper species **334** then should undergo the subsequent transmetallation step 2b more easily than the original tributyl tin residue. Step 2a can be further facilitated by addition of CsF[125] (see Table 5, entry 8). CsF acts as a source of fluorine, which either forms of a vinyl-tin ate-complex (increasing its leaving properties), or it simply removes the tributyltin iodide from the reaction by the formation of polymeric tributyltin fluoride. However, in our case, addition of copper salts did not lead to improved yields. According to Farina *et al.*,[126] with the proper choice of Pd-ligands the influence of copper salts is negligible. Actually, increased yields were found when tris-(*o*-furyl) phosphine or triphenylarsine ligands are applied without additives. Possibly, a competing isomerisation of the intermediate vinyl-copper species is responsible for this. On the other hand, application of the equally successful combination of LiCl/PPh$_3$[128] led to no isolable product after work-up, although full conversion could be observed by reaction control *via* TLC. We reasoned that a component in the reaction mixture could be responsible for this product destruction: Previously, the reactions were carried out in highly polar amide solvents like NMP or DMF – Bu$_3$SnI produced in the reaction could therefore coordinate or even react with the strongly coordinating solvent molecules present in abundance. This situation changes dramatically when the solvent is changed for THF – here, the most Lewis-basic position is the macrolactone carbonyl of product **224**, which most likely gets destroyed by this undesired Lewis-acid catalysis. This unwanted Bu$_3$SnI can be trapped *in situ* by addition of a scavenger to the reaction mixture. The addition of the organic salt *n*-Bu$_4^+$Ph$_2$PO$_2^-$ led to the formation of a precipitate (Ph$_2$PO$_2$SnBu$_3$) in the reaction mixture, and this time indeed, the product could be isolated in comparably good yield.

This elaboration shows that the anticipated intramolecular Stille-coupling had more than one pitfall, which all had to be determined experimentally and also circumvented by proper choice of reaction conditions:

- The ring size formed seems to be not very favourable – the high degree of unsaturation combined with the preferred *s-trans* configuration of the ester in the ground state move the reaction partners far apart from each other.

- The addition of copper salts has no beneficial effect in this case – a possible explanation for this observation is the isomerisation of the intermediate vinyl copper species.
- The reaction by-product Bu_3SnI can react with the reaction product in less polar solvents – trapping of this by-product is imperative to obtain good yields.

The achievement of the overall optimisation work can be summarised as follows: (Scheme 96) Inititally, macrolactone **224** was accessible from aldehyde **207** in 7 steps and 7 % overall yield. With the optimised procedures and strategies discussed in the preceding chapters, the synthesis of chiral hexalin **225** from aldehyde **207** is now possible in 5 steps and 19 % overall yield. When the sequences up to macrolactone **224** are compared, the optimisation resulted in a more than three-fold increase in yield.

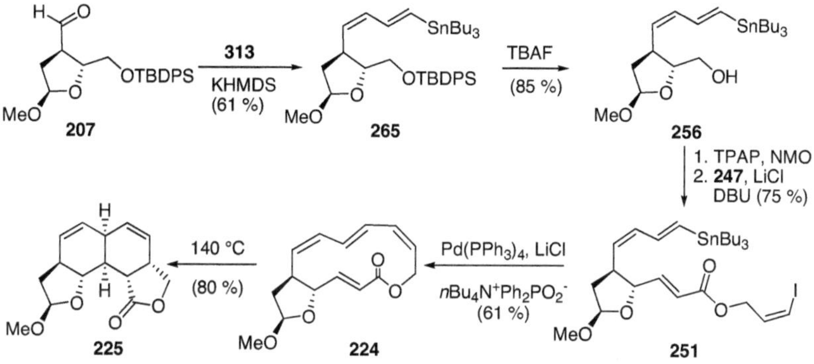

Scheme 96: Optimised synthesis of macrolactone **224** (summary)

7.4   *cis*-Hexalin functionalisation

With all problems associated with the preparation of **225** solved, we could approach its functionalisation. (Scheme 97) One of the remaining tasks was the replacement of the lactone–carbonyl carbon at C-10 with a hydroxy function. Then, the selective epoxidation of the C-7 – C-8 double bond directed by the C-21 hydroxy group should be possible. Finally, after methylation of the C-21 primary hydroxyl group, opening of the lactol ring would lead to one primary and one secondary alcohol. Selective protection of the former, followed by oxidation of the remaining secondary alcohol to the ketone should yield the fully functionalised *cis*-octalin **105**, ready for addition of the metallated side-chain.

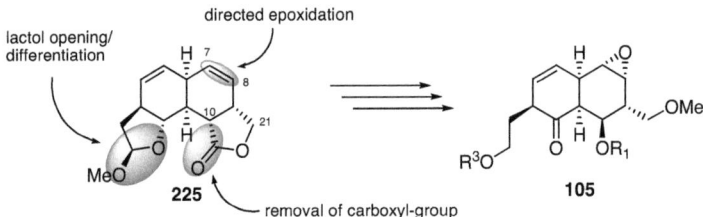

Scheme 97: Remaining functionalisation steps

Removal of the C-10 carbonyl carbon can be achieved most efficiently in a 3-step sequence, starting with an α-hydroxylation of the carbonyl, followed by reduction to the triol and a periodate cleavage of the 1,2-diol the yield the C-10 oxo-functionality. Therefore, the initial efforts focussed on the α-oxidation and/or opening of the lactone ring. (Scheme 98)

Unfortunately, the treatment of the lithium enolate of **225** with an oxygen-saturated solution of P(OMe)$_3$ in THF[130] did not lead to the C-10 oxidised product **337**. As this α-oxidation could be problematic due to the ring strain of the tetracyclic lactone, we alternatively thought to open the lactone first, *e.g.* by formation of the Weinreb amide **338**. Regrettably, treatment with *N,O*-dimethylhydroxylamine hydrochloride and isopropyl magnesium chloride[131] did not lead to the formation of **338**. To try something more conventional, we hoped at least the lactone opening should be possible with secondary amines like dimethyl amine or pyrrolidine.[132] But again, treatment with

---

[130] a) Corey, E. J.; Ensley, H. E. *J. Am. Chem. Soc.* **1975**, *97*, 6908 – 6909; b) Takeda, K.; Shibata, Y.; Sagawa, Y.; Urahata, M.; Funaki, K.; Hori, K.; Sasahara, H.; Yoshii, E. *J. Org. Chem.* **1985**, *50*, 4673 - 4681; c) Rath, J.-P.; Kinast, S.; Maier, M. E. *Org. Lett.* **2005**, *7*, 3089 - 3092.

[131] Williams, J. M.; Jobson, R. B.; Yasuda, N.; Marchesini, G.; Dolling, U.-H.; Grabowski, E. J. J. *Tetrahedron Lett.* **1995**, *36*, 5461 - 5464.

[132] Mulzer, J.; Giester, G.; Gilbert, M. *Helv. Chim. Acta* **2005**, *88*, 1560 - 1579.

Me₂NH at room temperature, at reflux or even in the microwave reactor at the highest possible temperature (160 °C for EtOH/Me₂NH) did not give any product **339**. The alternative opening with pyrrolidine to give amide **340** failed as well.

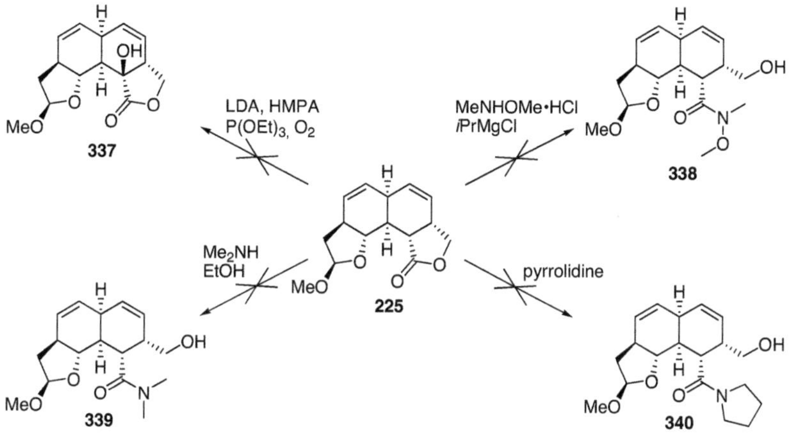

Scheme 98: Attempts to open the γ-lactone

In the former attempts, despite no conversion to a new product, the starting material could not always be reisolated. We therefore assumed that the cyclic methyl lactol in the tetracycle **225** might not be as stable as before the TADA reaction, maybe due to the strain imposed by the *cis*-hexalin ring. However, this situation could be defused by deliberately opening the lactol ring and protecting the two hydroxy functions orthogonally. Acetal hydrolysis with aqueous HCl and subsequent reduction of the free lactol with NaBH₄ gave the diol **341** in excellent yield (Scheme 99) The primary hydroxy function was selectively protected with the sturdy TBDPS group (**342**). As the secondary alcohol protecting group should be more easily removable, ideally under the oxidation conditions (Swern-oxidation!), we decided to protect it as a TES-ether to give fully protected *cis*-hexalin **343**.

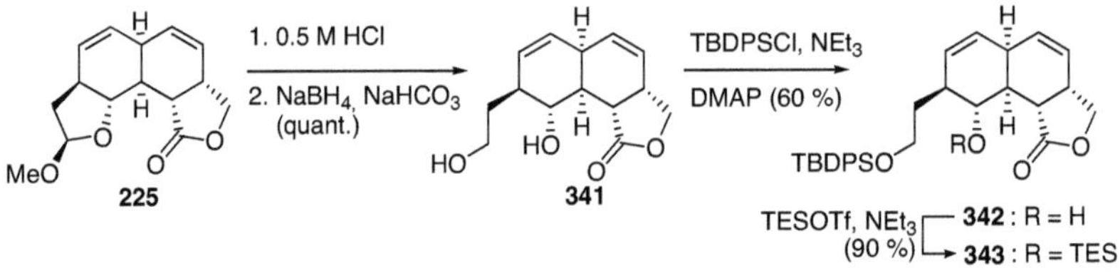

Scheme 99: Opening and differentiation of the lactol moiety

Now the opening of the γ-lactone was tried again, starting with the strategically ideal α-oxidation. (Scheme 100) This time, when the Li-enolate was treated with O_2 in the presence of $P(OMe)_3$, the starting material dissappeared, although without the concurrent appearance of a new product (like **344**). The same results were obtained when KO*t*Bu was used instead of LDA as a base. Also, formation of the TMS-enolate, followed by treatment with dimethyldioxirane (DMDO) did not give any product. We were again puzzled by this observation, as very similar molecules, lacking only the second cyclohexene ring annulated to the tetrahydroisobenzofuran-2-one, could be transformed in high yields to the desired α-hydroxy compound! In the end, we attributed the observed destruction of **343** to an oxy-ene reaction between the bridgehead proton and molecular oxygen in the solution, leading to a conjugated hydroperoxide. Therefore, the lactone had to be opened first and one double bond oxidised in a directed epoxidation. An initially attempted DIBAl-H reduction to the lactol **345** only led to traces of product, despite excess of reagent and long reaction times. Also an opening of the lactone with sodium methoxide to **346** was not successful. Eventually, opening of the lactone to the Weinreb amide under Williams' conditions[131] worked in excellent yields. Interestingly, upon lactone-opening, the proton signals of **347** in the NMR became less resolved – this indicates that by opening of the lactone, the system regained additional degrees of flexibility.

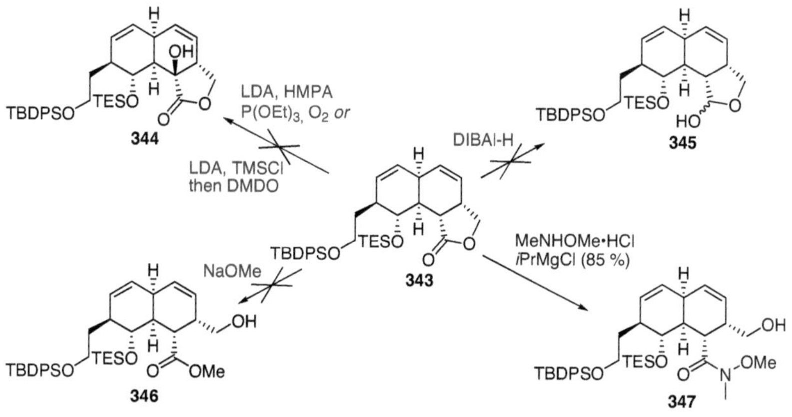

Scheme 100: Attempts to open the γ-lactone **343**

Finally, having achieved the opening of the lactone ring, the stage was now set to undertake the pivotal face-selective epoxidation[133] of the C-7 – C-8 double bond directed by the C-21 OH group. (Scheme 101) We first attempted the directed epoxidation using VO(acac)$_2$ as a catalyst,[134] as this system gives good results especially in the case of homoallylic alcohols. However, upon subjection of **347** to the reaction conditions, only small amounts of a new product were observed. When equimolar amounts of VO(acac)$_2$ and a large excess of *tert*-butylhydroperoxide (TBHP) were applied, full conversion was observed on TLC. Unfortunately, this turned out to be rather due to destruction of the starting material – only a little amount of the desired product **348** could be isolated. Closer inspection of the literature revealed that VO(acac)$_2$ directed epoxidations failed when the directing hydroxy-group was in a pseudo-equatorial position.[135] Although this rule was put forward only for ring-hydroxyl-groups, the same behaviour was observed in our case. To our delight, the directed epoxidation using *m*CPBA gave **348** as a single product, which we assumed to be the desired one. A first attempt to protect the primary alcohol under basic conditions

---

[133] For a review, see: Hoveyda, A. H.; Evans, D. A.; Fu, G. C. *Chem. Rev.* **1993**, *93*, 1307 – 1370.
[134] Sharpless, K. B.; Michaelson, R. C. *J. Am. Chem. Soc.* **1973**, *95*, 6136 – 6137.
[135] Gadwood, R. C.; Lett, R. M.; Wissinger, J. E. *J. Am. Chem. Soc.* **1986**, *108*, 6343 – 6350.

gave the desired methyl ether **349**. However, upon scale-up, the lactone closed again directly after the epoxidation. Such behaviour has been reported for similar systems.[86] Although undesired, epoxy-lactone **350** showed again well-resolved signals, which now allowed determination of the regio- and diastereocontrol of the directed epoxidation.

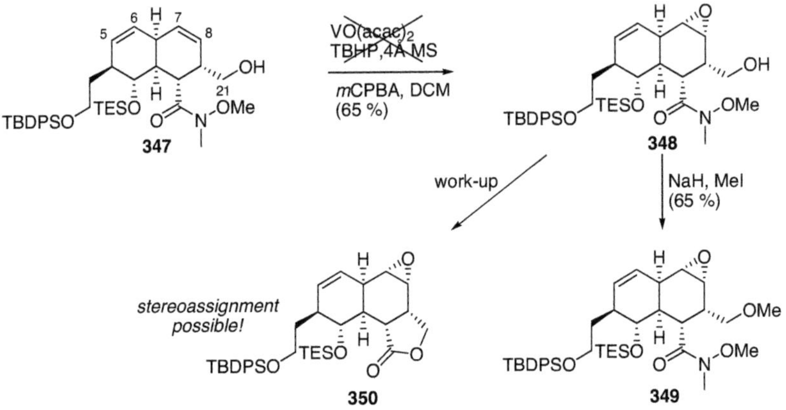

Scheme 101: Synthesis of **349** *via* hydroxy-directed epoxidation

The ^{1}H NMR of compound **350** now shows only two olefinic protons, which were assigned to be the A-ring protons H-4 and H-5. (Figure 31) Concomitant to the disappearance of the two olefinic protons, two additional signals have appeared at 3.23 and 3.09 ppm.

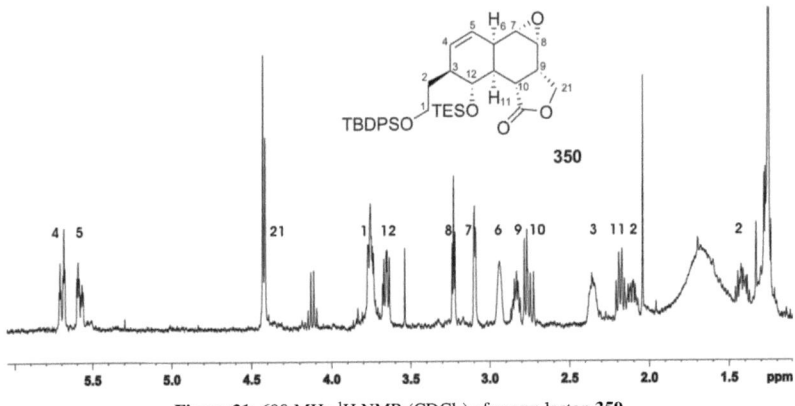

Figure 31: 600 MHz ^{1}H NMR (CDCl$_3$) of epoxy-lacton **350**

In contrast to previous NOE studies, the axial proton H-12 now shows transannular interactions with the epoxide proton H-7, besides the cross peak with proton H-9 already observed previously. (Figure 32) The latter NOE interaction shows that the *cis*-octalin still adopts the same conformation as previously seen on the *cis*-hexalin **225**. (Figure 33) Due to the additional transannular interaction between H-12 and H-7, the epoxide should indeed be *syn* to the C-21 hydroxymethylene group. As this epoxide corresponds to an attack of the peracid on the convex side of the molecule, structure **350** is not only proofed by the NMR data, but also the most likely based on the steric properties of the preceding *cis*-hexalin **347**.

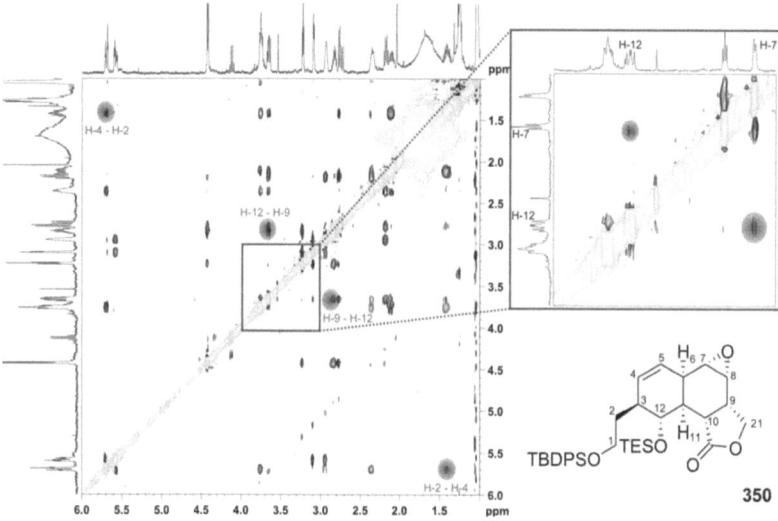

Figure 32: NOESY spectrum of **350** showing the key NOE interactions H-12 – H-7 and H-9 – H-12

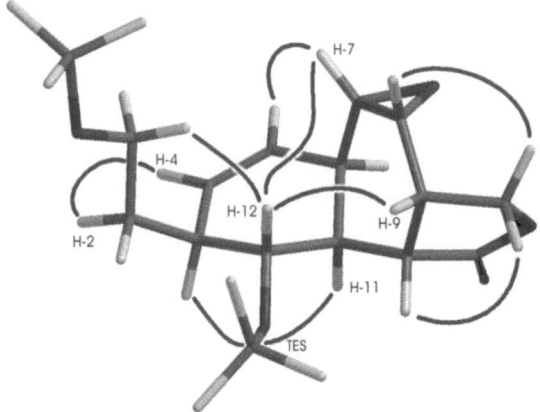

Figure 33: NOE-Interactions depicted on simplified 3D model of **350**

Unfortunately, re-opening of epoxylactone **350** to the Weinreb amide **348** could be achieved only in low yields, and the subsequent methylation again gave inferior yields. These transformations were carried out with the last material available. Having no further reserves, the synthesis had to be stopped here.

To accomplish the last key transformation necessary for the synthesis of the core fragment, the removal of the ester group, 3 further steps would have been necessary (Scheme 102). α-Oxidation of the Weinreb-amide **349**, followed by LiAlH$_4$ reduction will lead to diol **351** – the redundant carbon can then be removed by periodate cleavage. Reduction of the produced ketone should occur from the convex face, giving the *endo*-configured C-10 hydroxy group, which then could be protected as SEM-ether **352**. The SEM-group would have two advantages in this position – first, it is a sturdy and sterically undemanding protecting group, and second, it can be removed in a final, global deprotection step with a fluorine source (*vide infra*). Selective cleavage of the TES-group followed by a Swern-oxidation (probably possible in a one-pot procedure), finalises the synthesis of the desired, fully functionalised *cis*-octalin **336**.

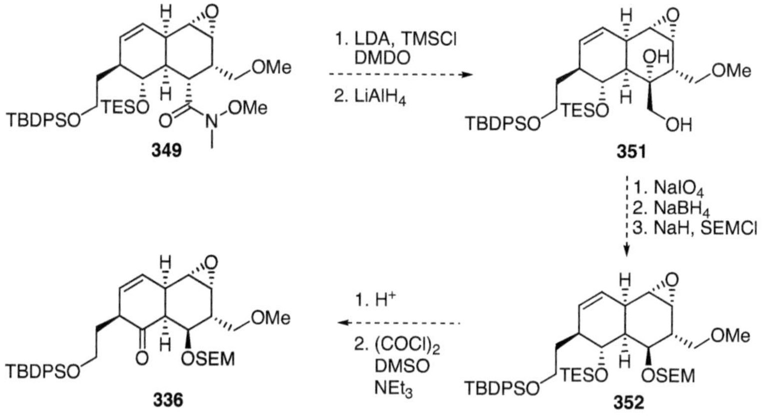

Scheme 102: Remaining steps to acess *cis*-octalin **336**

In summary, methyl protected *cis*-octalin **349** could be synthesised in 2.0 % overall yield in 13 steps starting from chiral butenolide **151**, or 18 steps (1.3 % overall yield) from (+)-ascorbic acid. The remaining conversion to the fully functionalised *cis*-octalin **349** ready for the attachment of the side chain would have taken another 7 steps, making the whole sequence account for 20 steps starting from butenolide **151**, or 25 steps starting from (+)-ascorbic acid (**92**).

8 SYNTHESIS OF THE C-13 – C-18 SIDE CHAIN

8.1 Concept

The strategy for the total synthesis of branimycin (**14**) is based on its retrosynthetic disconnection into the fragment **336** (its synthesis being discussed in detail in chapters 5-7) and the C-13 – C-18 side chain **106**. (Scheme 103) This side chain **106** should be attached to the core fragment ketone **336** in a 1,2-addition under concomitant transannular oxo-bridge formation and epoxide opening.

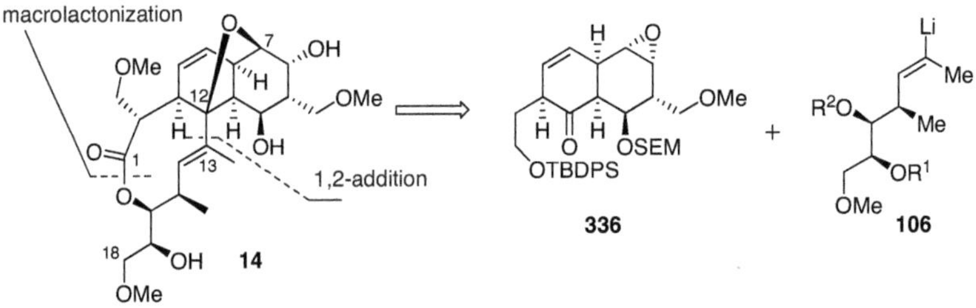

Scheme 103: Retrosynthetic disconnection into core fragment **336** and side chain **106**

The precursor for **106** is vinyl iodide **355**, which in turn could be easily obtained from alkyne **356**. (Scheme 104) In **356**, protecting group R^2 should be removed selectively prior to the macrolactonisation on an already densely functionalised molecule and R^1 should be sturdy enough to stay intact up to the final deprotection at the end of the synthesis. We thought the synthesis of this *syn*, *syn* stereotriade could be achieved the fastest by a disconnection between C-15 and C-16. This in turn gives protected glyceraldehyde derivative **357** and nucleophile **358** as synthons. To introduce the chiral C-15 methyl group, this reaction should be carried out using a chiral nucleophile like **358**. This proposition now raises the question of a double stereodifferentiation and a possible mismatch in the C-C bond forming step.

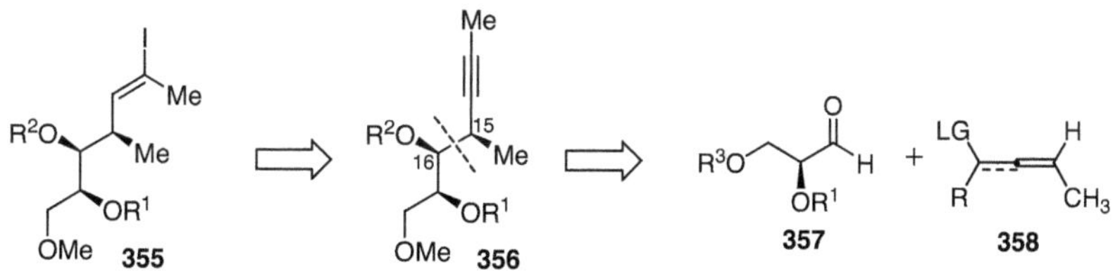

RESULTS AND DISCUSSION – C-13 – C-18 SIDE CHAIN SYNTHESIS

Scheme 104: Side chain construction via diastereoselective electrophilic substitution

8.2 Prior work

My colleague Daniela Rosenbeiger developed in her diploma thesis[136] an approach towards the side chain fragment **356** using a Roush-crotylation reaction (Scheme 105). As these investigations have some important implications on this work, they are described here briefly.

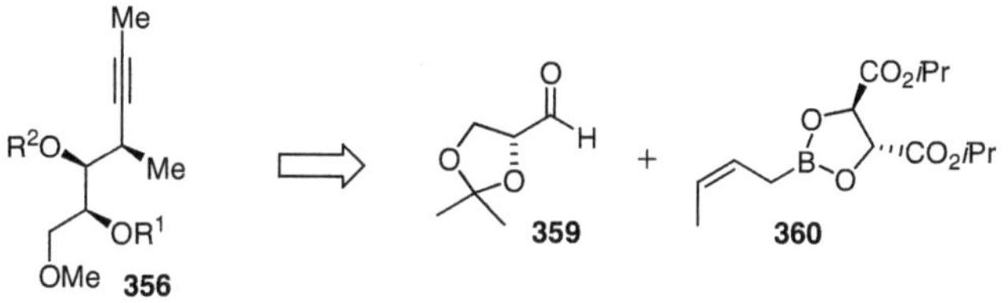

Scheme 105: Side chain construction *via* Roush-crotylation

Acetonide protected (*R*)-glyceraldehyde **359** was synthesised in 2 steps from D-mannitol. (Scheme 106) Chiral Z-butenyl boronate **360** was accessed by deprotonation of Z-butene with BuLi/KO*t*Bu, followed by addition of B(O*i*Pr)₃ – a transesterification with (*R,R*)-diisopropyltartrate produced the required Z-butenyl boronate **360**. (Scheme 106)

The reaction between **359** and **360** is known to go through a Zimmerman–Traxler transition state controlled by the chiral boronate ester, which in accordance with the Felkin-Anh direction of the chiral aldehyde makes the *anti,syn*-crotylation product **361**

---

[136] Rosenbeiger, D. *Diplomarbeit* Universität Wien, **2004**.

the matched product.[137] (Scheme 106) The secondary hydroxyl group in **361** was protected as its benzyl ether (the PMB ether turned out to be too unstable in the subsequent reaction sequence) and the acetonide hydrolysed to diol **362**. To obtain the desired *syn,syn* stereotriade, the secondary alcohol at C-17 had to be inverted, and the most efficient way to do this was *via* an epoxide ring formation/ring opening sequence: After selective benzoylation of the primary alcohol at C-18, followed by mesylation of the C-17 OH-function and subsequent treatment with NaOMe produced epoxide **363** under clean inversion of configuration. This detour was chosen as preliminary experiments had shown that a selective monomethylation of primary alcohol on the corresponding *syn,syn* triol (available in analogy, but less stereocontrolled) was not possible. The epoxide was opened at the primary position with an excess of NaOMe in MeOH and the C-17-hydroxyl group was silylated with TIPSOTf to give **364** in acceptable yields. Ozonolysis furnished the unstable aldehyde **365** which was immediately subjected to the Corey Fuchs alkynylation.[138] This reaction, though successful on a small scale, led to elimination of the benzyloxy-group when scaled up. To overcome this problem, conversion of the aldehyde **365** to the alkyne was achieved using trimethylsilyl diazomethane.[118] Methylation of the resulting alkyne finally led to compound **366**. Initial attempts to convert **366** into the *E*-vinylstannane **367** *via* a Pd-catalysed hydrostannylation failed, despite literature precedence.[139]

---

[137] Roush, W. R.; Hoong, L. K.; Palmer, M. A. J.; Straub, J. A.; Palkowitz, A. D. *J. Org. Chem.* **1990**, *55*, 4117-4126.

[138] Corey, E. J.; Fuchs, P. L. *Tetrahedron Lett.* **1972**, *36*, 3769 - 3772.

[139] Benechie, M; Skrydstrup, T.; Khuong-Huu, F. *Tetrahedron Lett.* **1991**, *32*, 7535 - 7538.

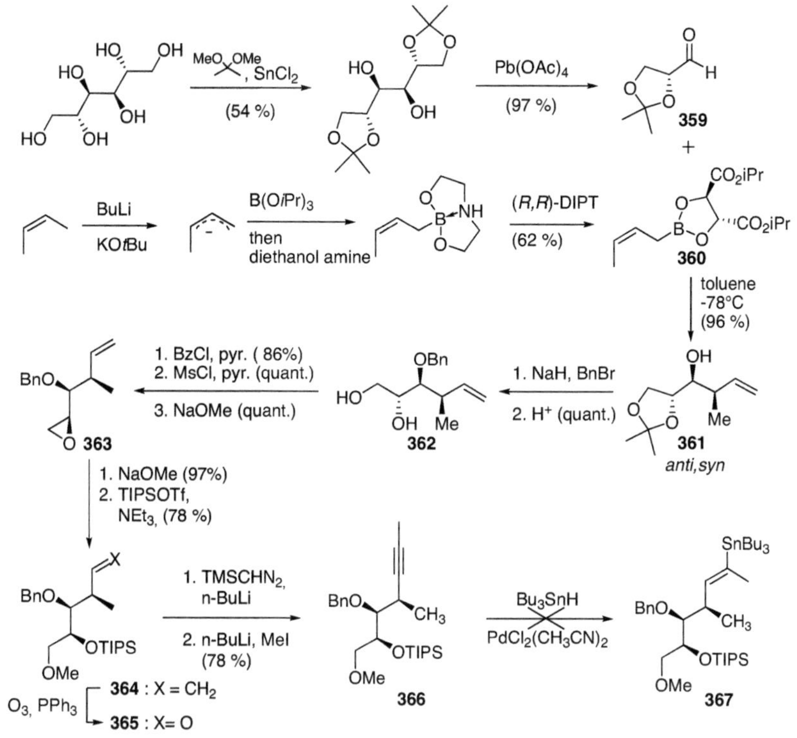

Scheme 106: Synthesis C-13 - C-18 side chain precursor **366** by D. Rosenbeiger[136]

Although this synthetic sequence furnished the side-chain precursor **366** in good overall yield (48 % over 11 steps), the sequence was linear and involved a significant number of reactions. Therefore we investigated a more convergent approach to access **366**.

8.3 Diastereoselective propargylation approach

On the basis of Marshall and Fleming's work on chiral allenylstannanes[140] and allenylsilanes[141] it was decided to introduce the C-13 – C-15 unit in a single step using an asymmetric propargylation reaction with allenyl silane **370**. (Scheme 107) To avoid too many functional group interconversions at later stages, chiral aldehyde **369** was designated to carry the OMe-group at C-18 already in place.

Interestingly, in stark contrast to the numerous mechanistic studies on diastereoselective propargylation with chiral allenylstannanes, only limited investigations have been reported for chiral allenylsilanes. Based on the experimental data available for allenylstannanes,[140a,b] the propargylation of an achiral aldehyde with allenylsilane **370** should proceed *via* an antiperiplanar transition state **370A** and lead to adduct **371**, which would have the desired and absolute configurations at C-4 and C-5. (Scheme 107)

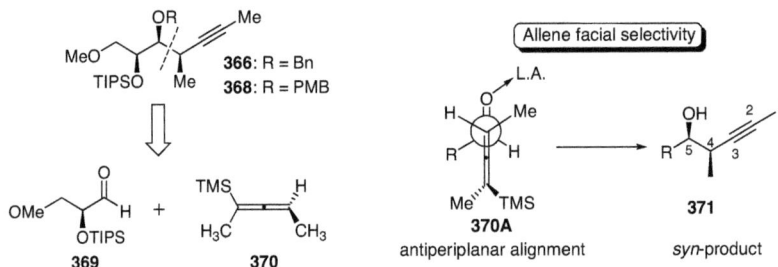

Scheme 107: Side chain construction *via* chiral allene propargylation

It had not yet been determined how the two oxygen substituents on the aldehyde would influence the trajectory of the incoming allenyl unit and thus affect the stereochemical outcome of the reaction.

---

[140] a) Marshall, J.A. *Chem Rev.* **1996**, *96*, 31 – 48; b) Marshall, J.A.; Wang, X. *J .Org. Chem.* **1992**, *57*, 1242 - 1252. c) Marshall, J.A.; Chobanian, H. *Organic Syntheses* **2005**, *82*, 43-50.

[141] a) Buckle, M. J. C.; Fleming, I. *Tetrahedron Lett.* **1993**, *34*, 2383 – 2386. b) Marshall, J. A.; Maxson, K. *J. Org. Chem.* **2000**, *65*, 630 – 633. c) Buckle, M. J. C., Fleming, I.; Gil, S.; Pang, K. L. C. *Org. Biomol. Chem.* **2004**, *2*, 749 – 769.

In an initial investigation, enantiomerically pure aldehyde **369** was prepared by a visiting student, Daniele Castagnolo, from (R,R)-dimethyltartrate **373**.[142] The diol was protected as the acetonide and was then reduced with LiAlH$_4$ to afford diol **374**. (Scheme 108) Under standard conditions (NaH, DMF, MeI) the mono-methyl ether **375** was formed predominantly and only small amounts of the di-methyl ether were detected. The primary alcohol in compound **375** was tosylated (**376**) and converted to its corresponding iodide **377**. Reductive elimination of the acetonide with activated Zn, furnished allylic alcohol **378**.[142] After TIPS-protection (now possible with TIPSCl), ozonolysis of **379** now afforded enantiomerically pure (S)-aldehyde **369**.

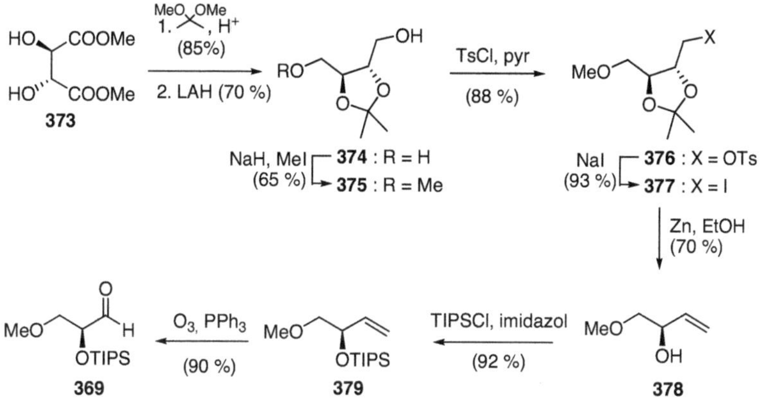

Scheme 108: Synthesis of chiral aldehyde **369** by D. Castagnolo

Alternatively, a much shorter route was employed starting from (S)-glycidol **380**, which was O-methylated using the soft base Ag$_2$O and MeI as a methylating agent. (Scheme 109) These conditions were crucial for the success of the reaction, as harsher conditions (NaH, MeI) led to dimerisation of glycidol. Addition of a sulfonium ylide to **381** (produced by lithiation of trimethyl sulfonium iodide)[143] led to opening of the epoxide

---

[142] Rao, A. V. R.; Reddy, E. R.; Joshi, B. V.; Yadav, J. S. *Tetrahedron Lett.* **1987**, *28*, 6497 – 6500.
[143] Alcaraz, L.; Harnett, J. J.; Mioskowski, C.; Martel, J. P.; Le Gall, T.; Shin, D.-S.; Falck, J. R. *Tetrahedron Lett.* **1994**, *35*, 5449 – 5452.

to give the intermediate betaine **382**, which upon elimination of dimethylsulfide yielded allylic alcohol **378**.

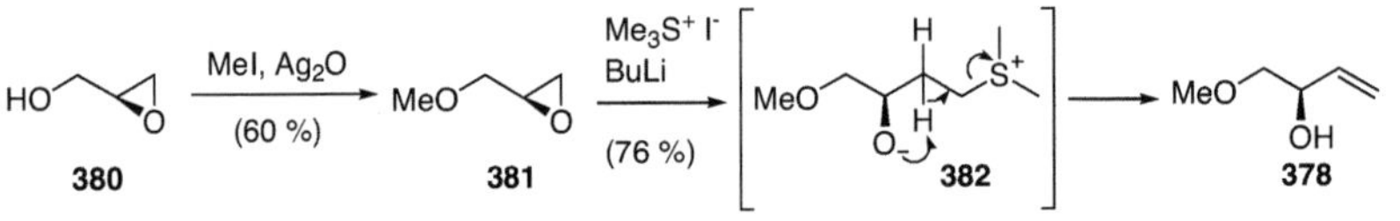

Scheme 109: Shortened synthesis of allylic alcohol **378**

Non-racemic allenylsilane **370** was synthesised starting from commercially available, racemic TMS-butyne-2-ol **383**.[141c,144] Lipase-catalysed chiral resolution of the alcohol led to formation of (*S*)-TMS-butyn-3-ol **384** in ≥98% ee,[144] along with equally enantiomerically pure (*R*)-propargyl acetate **385**. (*S*)-TMS-butyn-3-ol **384** was mesylated at -78 °C. After work-up, the crude mesylate **386** was treated with MeMgCl in the presence of equimolar amounts of copper bromide, resulting in a clean S$_N$2' reaction to yield chiral allene **370** in excellent yield.

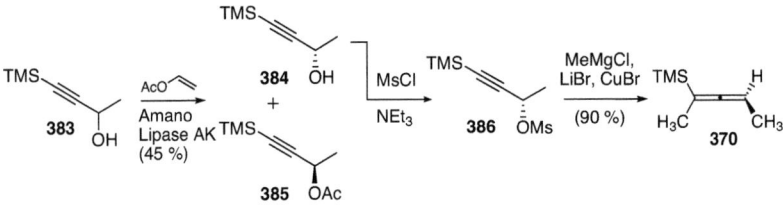

Scheme 110: Synthesis of chiral allenylsilane **370**[141c]

The addition of **370** to aldehyde **370** was mediated by TiCl$_4$, the preferred Lewis-acid in these allenyl silane addition reactions.[141] Under these conditions, a homopropargyl adduct was formed as a 20:1 diastereomeric mixture in 77 % yield. (Scheme 111)

---

[144] Bahadoor, A. B.; Flyer, A.; Micalizio, G. C. *J. Am. Chem. Soc.* **2005**, *127*, 3694 – 3695.

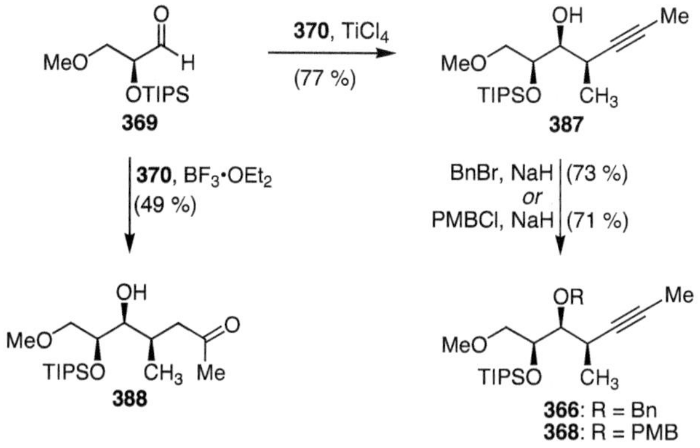

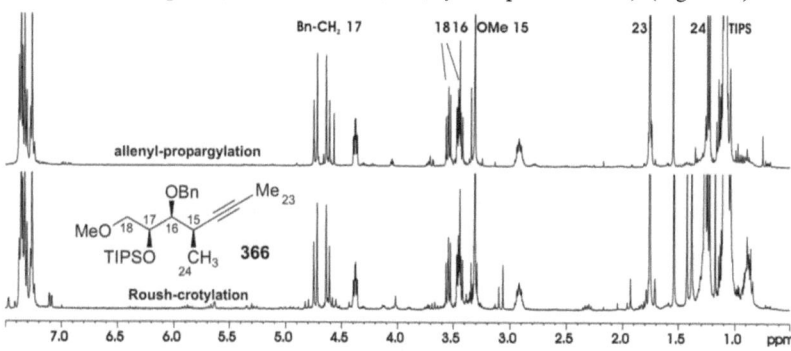

Scheme 111: Propargylation based synthesis of **366**

To our delight, after O-benzylation, the main diastereomer was found to be identical with **366** in all respects (^{1}H and ^{13}C NMR, MS, R_f and optical rotation). (Figure 34)

Figure 34: ^{1}H NMR of the *syn,syn* stereotriades derived from Roush crotylation and allenyl propargylation

To corroborate this conjecture, PMB-ether **368** was converted to cyclic acetal **389** and NOESY-experiments confirmed the proposed stereochemical assignment. (Scheme 112)

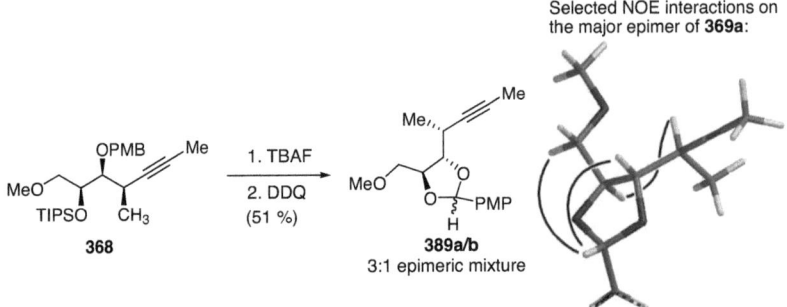

Scheme 112: Proof of *syn, syn* stereochemistry by acetal formation

The propargylation of **369** with allenyl silane **370** was also attempted using $BF_3 \cdot OEt_2$ as a non-chelating Lewis-acid. However, this reaction produced methyl ketone **388** in 49 % along with only a small amount of **387**. (Scheme 113) Interestingly, the diastereomeric excess of **387** turned out to be equally high in this case. This is surprising in the sense that $BF_3 \cdot OEt_2$ is a strictly non-chelating Lewis-acid, whereas $TiCl_4$ normally shows strong chelation. Further variations of the Lewis acids did not bring any improved results – the reaction with $TiBr_4$ worked well according to reaction control by TLC, although a much lower yield was isolated after work-up. When $SnCl_4$ was used as a Lewis-acid, besides the previously known products **387** and **388**, another new product was isolated, which was identified as **390** by HMQC and NOESY spectra. Finally, when $AlCl_3$ was used, besides **388** a complex mixture of products was obtained.

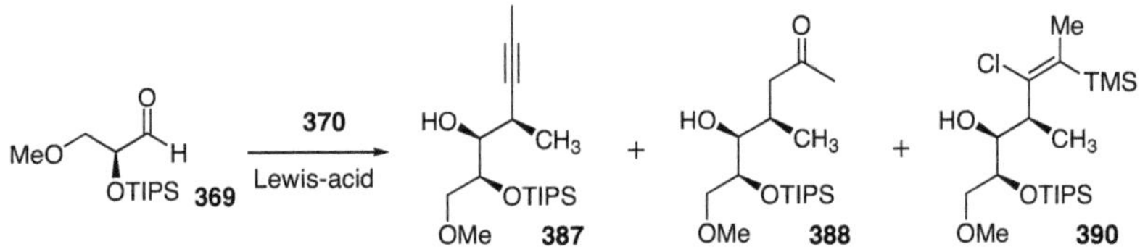

entry	Lewis-acid	d.r.	387	388	390
1	$TiCl_4$	20:1	77 %	20 %	-
2	$BF_3 \cdot OEt_2$	20:1	5 %	49 %	-
3	$TiBr_4$	n.d.	20 %	n.d.	-
4	$SnCl_4$	n.d.	9 %	30 %	5 %
5	$AlCl_3$	-	-	complex mixture	-

Scheme 113: Lewis-acid variation in the propargylation step

No clear explanation can be given about the predominant formation of the methyl ketone **388** in the presence of BF$_3$·OEt$_2$. What can be excluded is the addition of water to the triple bond after the propargylation reaction, as these reactions were executed under strictly anhydrous conditions. Reactions of allenyl silanes to give molecules related to **388** are known in the literature:[145] phenylpropanal (**391**), when treated with bissubstituted allene **392**, gave a 2:1 mixture of propargylation product **393** and dihydrofurane **394**.[145a] (Scheme 114) Acetaldehyd **395**, treated with the bulky allenyl TBS-silane **396**, produces the diastereomeric dihydrofuranes **397** and **398**.[145b] This 'side reaction' is a useful transformation in its own sense.[146] In our case, the observed methyl ketone **388** can be imagined as the hydrolysis product of the related 3,4-dihydro-2-trimethylsilyl-furane.

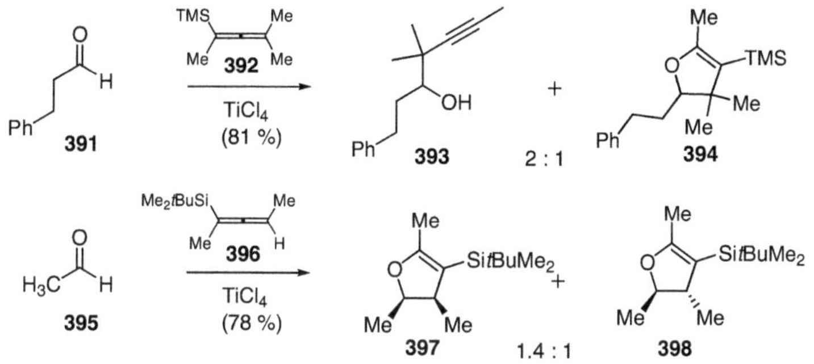

Scheme 114: Concurring 2+3 annulations observed by Danheiser *et al.*

When put together, these observations give a good insight into the possible reaction pathways: After activation of an aldehyde by a Lewis-acid, the allenyl silane acts as an electron-donor to form the intermediate vinyl-carbanion **399**. (Scheme 115) If the reaction follows the desired pathway (path A), the trimethylsilyl group is eliminated in an E$_1$-fashion to give the homopropargylic alcohol **400**. However, the same vinyl-carbanion **399** can also react with an external nucleophile (*e.g.* a halogen, delivered

---

[145] a) Danheiser, R. L.; Carini, D. J.; Kwasigroch, C. A. *J. Org. Chem.* **1986**, *51*, 3870 – 3878; b) Danheiser, R. L.; Kwasigroch, C. A.; Tsai, Y.-M. *J. Am. Chem.Soc.* **1985**, *107*, 7233 – 7235.

[146] For a review on this topic, see: Masse, C. E.; Panek, J. S. *Chem. Rev.* **1995**, *95*, 1293 – 1316.

either by the Lewis-acid or as a solvated anion) and form the vinyl halogenide **401** (path B). Given a sufficient half-life of vinyl carbanion **399**, it can also undergo a Si-1,2-shift to the isomeric vinyl carbanion **402**, which is now in a 1,5 relationsship with the newly formed alcohol (path C). Besides an E$_1$-elimination of the silyl group to give again the homopropargylic product **400**, the oxygen can attack the carbanion to give the dihydrofuran species **403**. Normally, the preference between path A and path C can be tuned by the size of the silyl group. However, when very bulky aldehydes like **369** are employed, this does not seem to be true anymore.

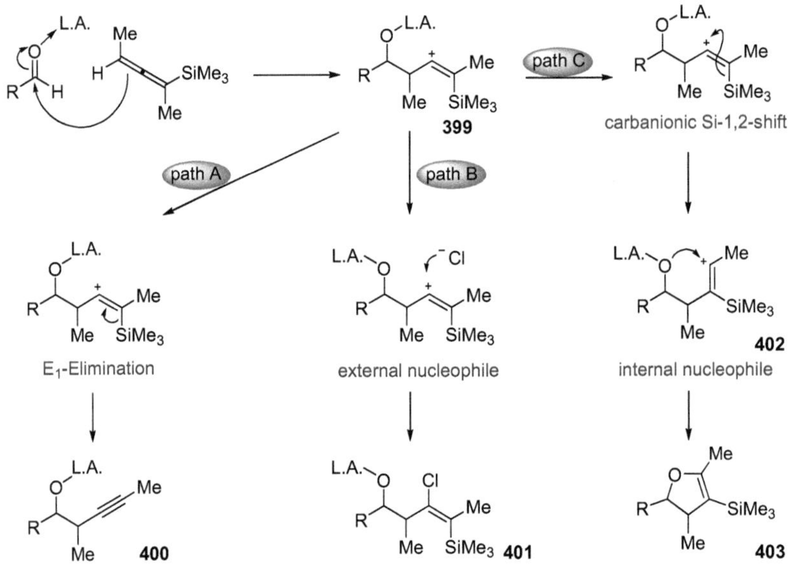

Scheme 115: Possible reaction pathways in allenyl silane electrophile addition

The observed stereochemical outcome of the addition is the result of an allenyl-*re*-aldehyde-*si*-face combination. (Figure 35) For such allenylsilane-aldehyde additions, two transition states **387A** and **387B** are discussed in the literature. Normally, the antiperiplanar geometry **387A** is favored over its synclinal counterpart **387B**, and indeed, in our investigations **387A** was preferred. It is quite obvious that the C-3 carbon of the allenyl silane will attack the aldehyde from the less hindered (*i.e.* Me-substituted) *re*-face. Much less obvious is the *si*-face preference of the aldehyde. The corresponding transition state could be described as **387C** or **387E**, which are both non-conventional:

387C is an *anti*-Felkin-Anh geometry, and **387E** would imply an even less likely chelate formation to an OTIPS-group. It seems more likely that either a chelate transition state **387F**,[147] previously postulated for allenylstannane additions to α-benzyloxy β-methyl propanals,[141b] or a "standard" Felkin-Anh geometry **387D** with the bulky OTIPS groups in the perpendicular position would be expected. However, both these models would lead to a *re*-face attack at the aldehyde, which could not be proposed on the basis of the product distribution.

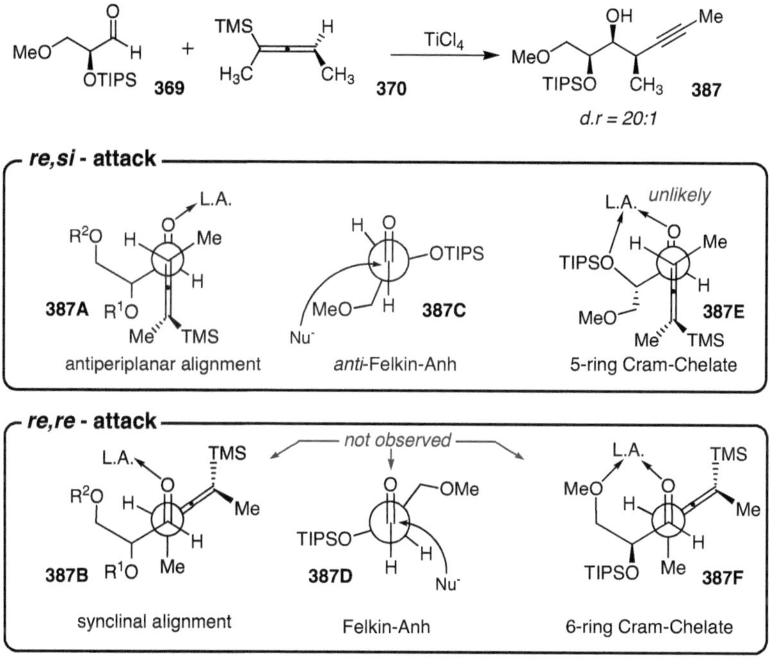

Figure 35: Rationale for the observed *syn,syn* selectivity

To get a better insight into the mode of activation of aldehyde **369**, we decided to investigate the complex of TiCl$_4$ and **369** at −78 °C by the means of deep-temperature

---

[147] Reetz, M. T. *Acc. Chem. Res.* **1993**, *26*, 462 – 468.

NMR spectroscopy. In a related investigation, Keck *et al.* could show that 6-ring chelates are formed from β-benzyloxy-α-methyl propanal and TiCl$_4$ at −78 °C.[148] In our case, this situation was not that clear. (Figure 36) At room temperature and upon cooling to −78 °C the ^{1}H NMR signals behaved as expected − upon addition of TiCl$_4$, a whole array of new signals formed. Analysis of this complex mixture by HH-COSY and ROESY spectroscopy could show, that at least 3 distinctly different species were present at −78 °C − one of them, signified by the aldehyde peak shifted 0.3 ppm to the upfield, could be attributed to the 6-ring chelate **369A**. Based on the ROESY signals, this peak is in exchange with two further signals at 6.82 and 6.25 ppm. At least one of these peaks might be attributed to *s-trans* complex **369B**. However, a direct comparison with literature data[149] was not possible, as ^{13}C measurements were not possible at -78 °C. The integration of these 3 signals indicated that the 6-ring chelate is only the minor component in this mixture. From the COSY, the OTIPS-methine protons of 2 conformers could also be identified. Unfortunately, no structural information of these conformers was available from the ROESY.

---

[148] Keck, G. E.; Castellino, S. *J. Am. Chem. Soc.* **1986**, *108*, 3847 − 3849.

[149] For ^{13}C-studies on BF$_3$-carbonyl complexes, see: a) Hartman, J. S.; Stilbs, P. *Tetrahedron Lett.* **1975**, *16*, 3497 − 3500; b) Torri, J.; Azzaro, M. *Bull. Soc. Chim. Fr.* **1978**, 283 − 291.

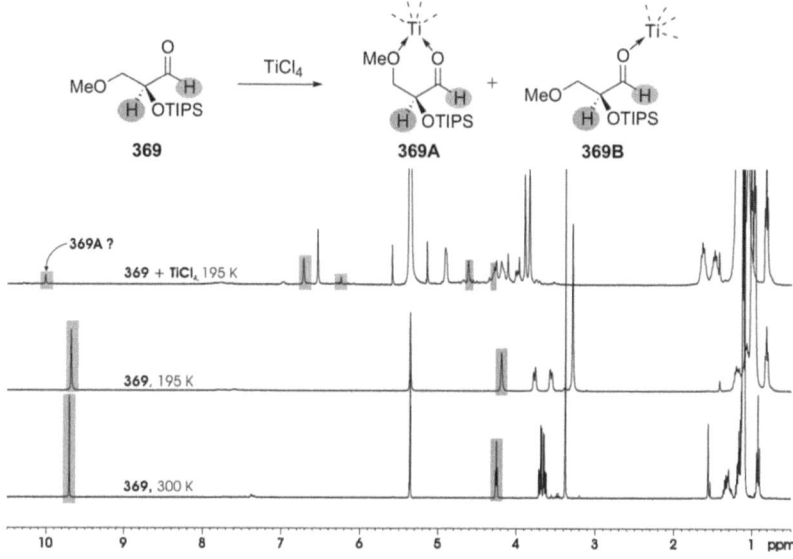

Figure 36: Deep-temperature NMR of aldehyde **369**-TiCl₄ complex

As not much information could be derived from the NMR-analysis of the Lewis-acid complex at deep temperature, we decided to get at least further insight into this reaction on a macroscopic level. Therefore, allene **370** was treated with the enantiomeric (*R*)-glyceraldehyde derivative **404**.[150] (Scheme 116) If the reagent still prefers transition state **370A**, then adduct **406** (*anti*-diol configuration) should be formed in excess. In reality, the reaction yielded compounds **405** : **406** in a ratio of 7 : 3. After separation, the relative configuration of each diastereomer was assigned *via* the cyclic acetals **407** and **408**, respectively.

---

[150] Prepared in analogy from (*R*)-glycidol – *cf*. Scheme 109

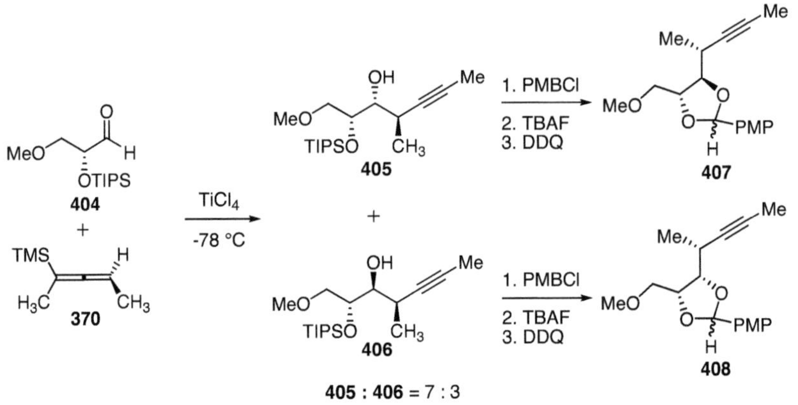

Scheme 116: Propargylation with enantiomeric aldehyde **404** and structural assignment of products

These results indicate that aldehyde **404** prefers conformation **404C** for the nucleophilic attack and thus, the allenyl component must switch from the antiperiplanar alignment **404A** to the synclinal arrangement found in **404B**. (Figure 37) Consequently, we postulate that for our system a combination of **369** and **370** represents the matched pair and compounds **404** and **370** represent the mismatched one. In the mismatched combination the aldehyde is the slightly dominating partner.

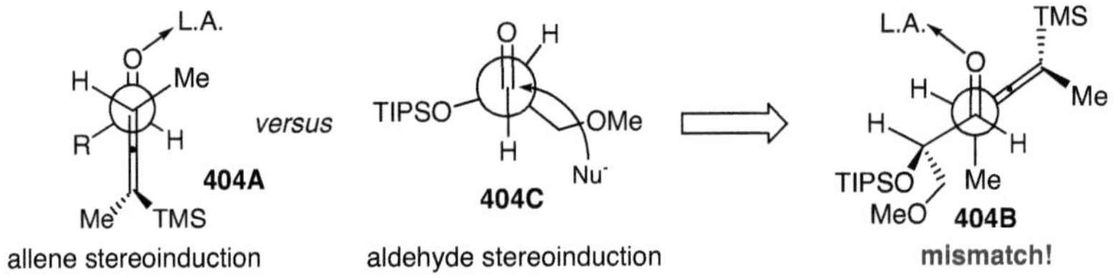

Figure 37: Mismatched combination of transition states

With this, we were able to show that allenyl silane additions to glyceraldehyde derivatives can be performed with high stereocontrol; however the diastereofacial induction exhibited by the aldehyde[151] does not fit into the standard transition state models.[152]

After these extensive studies, we returned to the overall synthesis of the C-13 – C-18 branimycin side chain. Therefore, **368** was subjected to a hydrozirconation (the Schwartz reagent was prepared by *in situ* reduction of Cp_2ZrCl_2 with DIBAl-H)[153] which, upon treatment with iodine, gave the E-vinyliodide **410**. (Scheme 117)

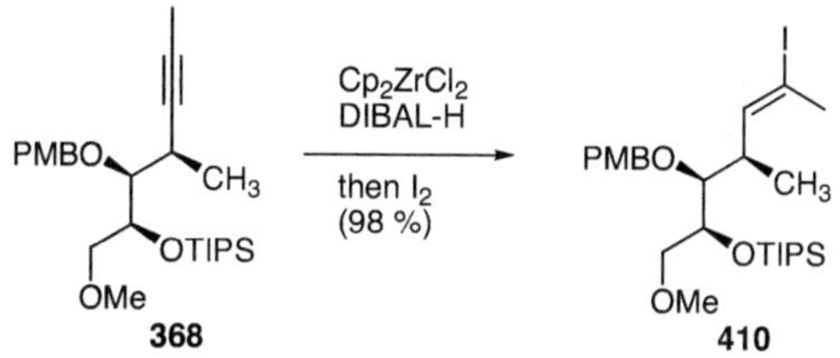

Scheme 117: Hydrometallation/halogen exchange to give vinyl iodide **410**

Having now a fully functionalised side-chain **410** in hands, it was now imperative to assay if a metallated side-chain **411** would now add to a *cis*-octalin carbonyl. For this purpose, the reaction with racemic, but otherwise fully functionalised *cis*-octalin **412** (provided by Dr. Valentin Enev) was investigated. (Scheme 118) At −78 °C, the side chain **410** was treated with 2 equivalents of *t*-BuLi; after addition of 3 equivalents of LiCl and octalin ketone **412** was added and, after 30 minutes, the reaction was quenched. TLC analysis showed the formation of two new spots, which could be attributed to the diastereomeric octalins **413** or **414**, based on the presence of both side-chain and octalin signals, although without the epoxide signals in the ^{1}H NMR.

---

[151] For other protecting group induced transition state changes in related transformation, see: a) Mikami, K.; Matsukawa, S.; Sawa, E.; Harada, A.; Koga, N. *Tetrahedron Lett.* **1997**, *38*, 1951 – 1954; b) Figueras, S.; Martín, R.; Romea, P.; Urpí, F.; Vilarrasa, J. *Tetrahedron Lett.* **1997**, *38*, 1637 – 1640.

[152] For examples where these rules are followed see: a) Williams, D. R.; Wultz, M. W. *J. Am. Chem. Soc.* **2005**, *127*, 14550 – 14551; b) Reymond, S.; Cossy, J. *Eur. J. Org. Chem.* **2006**, 4800 – 4804.

[153] Huang, Z; Negishi, E.-i. *Org. Lett.* **2006**, *8*, 3675 – 3678.

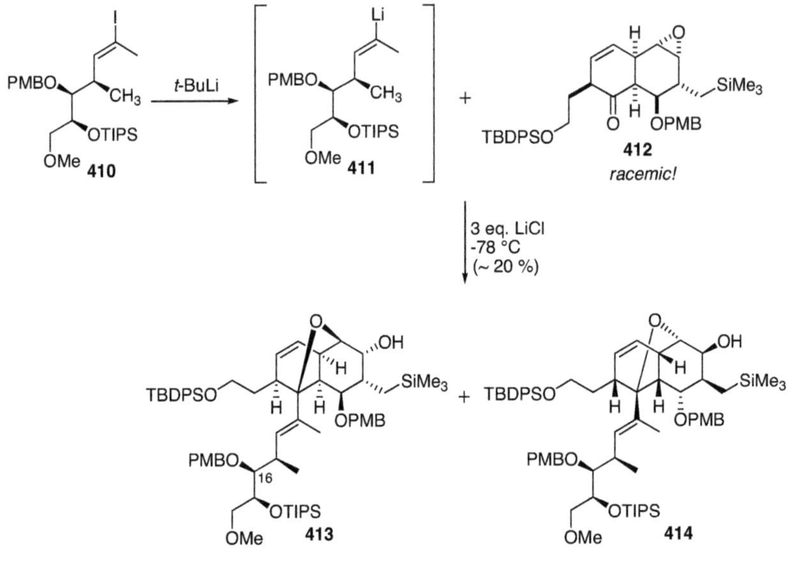

Scheme 118: Tested coupling of fragments **410** and **412**

Using this newly developed diastereoselective propargylation reaction, the synthesis of the side-chain fragment was achieved in 7 steps and an overall yield of 21 %. The protecting group at C-16 OH-group can be varied rather broadly to allow its orthogonal liberation later in the synthesis. Furthermore, it was also shown that side-chain fragment **410** can indeed act as nucleophile and add to a *cis*-octalin carbonyl in the desired manner.

9 CONCLUSION AND OUTLOOK

In this thesis, various synthetic approaches to both key building blocks of the antibiotic branimycin have been discussed. The synthesis of the *cis*-octalin core fragment had to be left short of completion; on the other hand, a very efficient synthesis of the C-13 – C-18 was found. The results are summarised and reviewed in the following chapters.

9.1 *Cis*-octalin core fragment

After initial attempts to use a D-ribonolactone derived chiral tether for an intramolecular Diels-Alder reaction (chapter 5), the strategy was changed to use a chiral γ-lactol instead for the chiral induction. (chapters 6 & 7) This chiral lactol was derived from L-ascorbic acid (**92**); here, a 4 step sequence led to the chiral, Z-α,β unsaturated ester **147**. (Scheme 119) Acid-catalysed lactonisation followed by TBPDS protection gave the chiral butenolide **152**, the central building block, whose chirality could now be proliferated by subsequent substrate controlled steps. After copper-catalysed vinyl-1,4-addition followed by DIBAl-H reduction, the chiral 5-ring had to be protected as the methyl lactol **206a**. Ozonolysis converted the vinyl-group into aldehyde **207a**, which was olefinated in a Z-selective Julia olefination to give Z,E-dienylstannane **265**. After removal of the TBPDS-protecting group, the free alcohol **256** was oxidised with TPAP followed by a Roush-Masamune olefination with phosphonate **247** to yield the *seco*-compound **251**. In a highly optimised intramolecular Stille-coupling, the vinyl-iodide and vinyl-stannane ends of **251** could be coupled to give macrolactone **224**. This macrolactone underwent a thermal Diels-Alder reaction to give the desired *cis*-hexalin **225** in a 10:1 selectivity. After reductive opening of the lactol and selective protection of the two alcohols, lactone **343** could be opened to the Weinreb amide **347**. Hydroxyl-directed epoxidation led to exclusive formation of the desired epoxide; methylation of the primary alcohol then gave methyl ether **349**. As material had run out at this stage, the remaining 7 steps (see Scheme 102) could not be carried out.

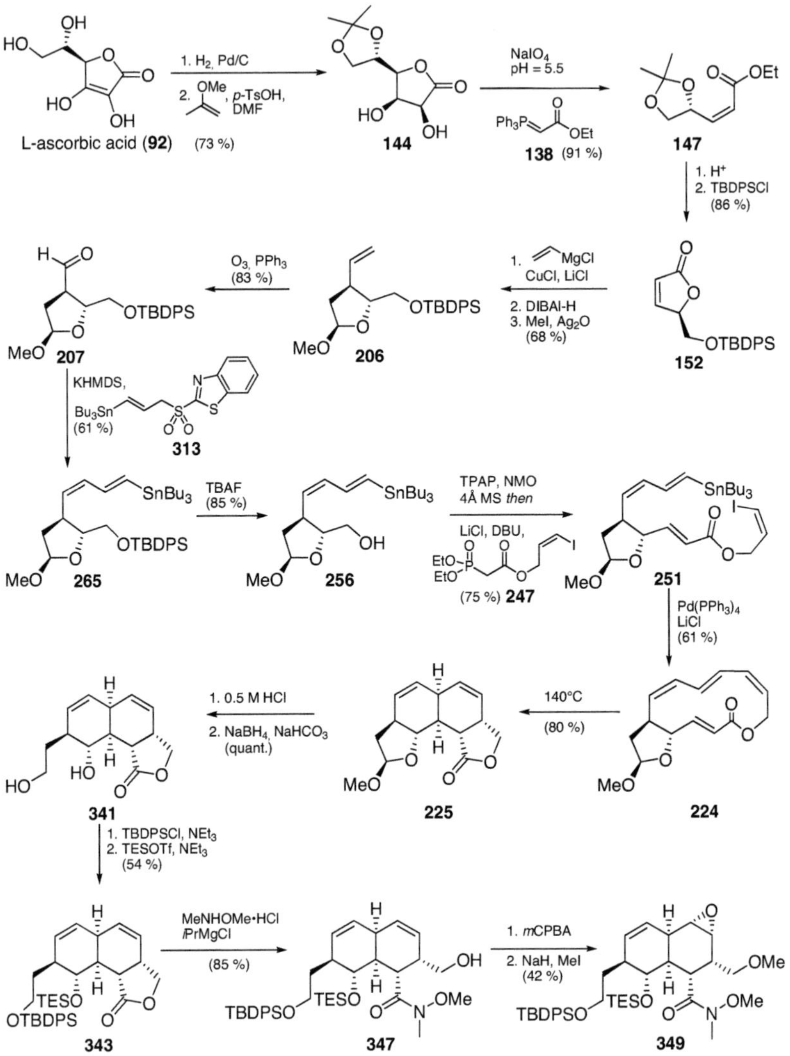

Scheme 119: Summarised *cis*-octalin synthesis

Although, disappointingly, the synthesis of the *cis*-octalin core fragment of branimycin could not be completed, the following results can be drawn from the synthesis carried out to this point.

1. Chiral, 4,5-*trans*-substituted lactols like **206a** can be used as efficient carriers of chiral information for further intramolecular transformations, especially Diels-Alder reactions.

2. The proliferation of chiral information in all transformations in this sequence can be efficiently realised exclusively under substrate control.

3. In the macrocyclic tetraene **224**, the Z,E,Z-triene unit can indeed act as a diene in a transsannular Diels-Alder reaction. In this reaction, the limited flexibility of the highly unsaturated macrolactone, combined with the conformational constraint of the chiral tether, forces the reaction to go predominantly through one *exo*-transition state.

4. When locked into a macrocycle, the Z,E,Z triene is more stable than its Z,E,E counterpart. Even upon prolonged heating, it maintains its configurational stability.

9.2 C-13 – C-18 Side Chain

In comparison to earlier work, a new strategy to access the side-chain fragment **368** in a very rapid manner was developed. Key concept in this strategy was the construction of the chiral homopropargylic alcohol **387** *via* addition of an axially chiral allene to an orthogonally protected glyceraldehyde derivative. (Chapter 8.3) This glyceraldehyde derivative was synthesised starting from (*S*)-glycidol (**380**), which was converted in a 2-step sequence to chiral allylic alcohol **378**. (Scheme 120) TIPS-protection and ozonolysis now delivered the required aldehyde **369**. The chiral nucleophile, allene **370**, was synthesised from propargylic alcohol **383** *via* kinetic resolution, mesylation and S_N2' displacement. These two fragments could be combined under Lewis-acid activation. The so produced *syn,syn* adduct **387** was protected as the PMB-ether **368**; hydrozirconation/iodination then delivered the fully functionalised side-chain **410**. After the lithiation of **410**, it could be added successfully to a test *cis*-octalin ketone **412** to give the desired oxo-briged octalin **413**.

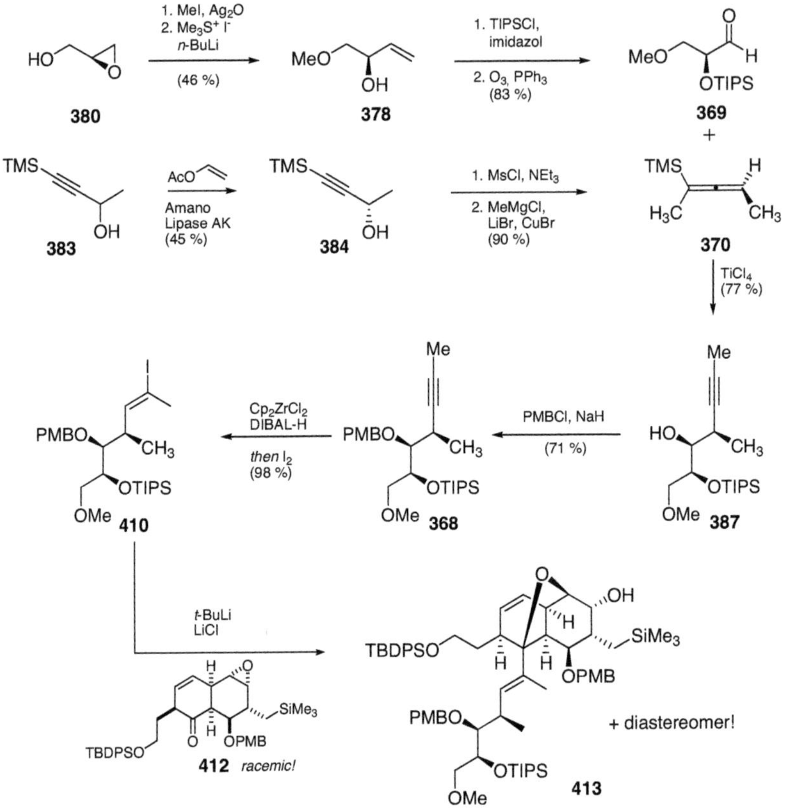

Scheme 120: Summarised side-chain synthesis and coupling

The described approach to access the branimycin C-13 – C-18 side chain *via* a diastereoselective propargylation reaction was used successfully to obtain multi-gram amounts of the desired building block in only 7 steps. In this convergent approach, it was found that the addition of chiral allene **370** to the very bulky glyceraldehyde derivative **369** did not follow the standard reaction pathways described for simpler systems. In this situation, the aldehyde **369** did not form a 6-ring chelate with the Lewis-acid – instead, an *anti* Felkin-Anh conformation was responsible for the reaction outcome.

9.3 Outlook / Possible improvements

Despite the extensive effort put into an efficient synthesis of the macrocycle **224** preceding the TADA-reaction, its synthesis is still flawed by three consecutive steps with an individual maximum yield of 61 to 75 %, which corresponds to a yield of 28 % over these steps. Considering the amount of steps following these transformations, they certainly represent a bottle neck. The fragmentation methodology described in chapter 7.2.5 provides a completely different approach to the same macrolactone **224**, although without the need of a Pd-catalysed cross-coupling or a semi-efficient Julia-olefination. Although briefly investigated, this quite promising approach could not be fully assessed. Some further investigations in this field might be appropriate to evaluate this strategy (Scheme 121). Besides the α-1,2-selective Reformatsky reaction on α,β-unsaturated aldehyde **294** proposed in chapter 7.2.5, another promising alternative is the use of α,β-unsaturated aldimine **290**. Imine **290** has the advantage of offering only one lone-pair for potential coordination to the metallated pyranone. In both cases, the aldol product can then be converted into the activated allyl tosylates **293** and **291**, which upon metallation, could fragment to give macrolactone **224**.

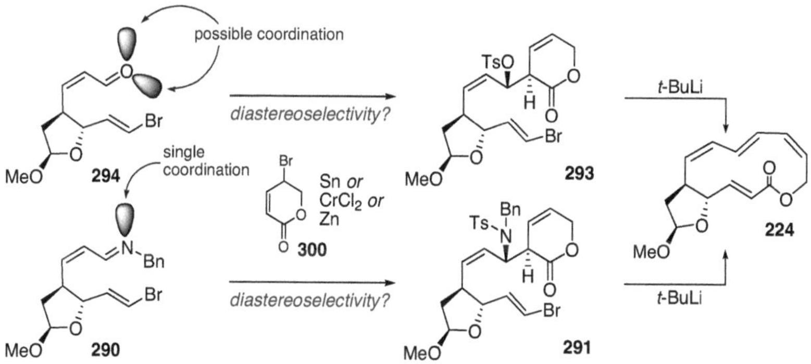

Scheme 121: Alternative fragmentation approaches

Also the intramolecular Stille-coupling still offers room for improvements. The difficult closure of the 13-membered ring might be due to the inherent preference of the *seco*-ester **251** for the ester *s-trans* configuration **251A**, which is about 9 kcal/mol lower in energy than the *s-cis* conformer **251B**. (Scheme 122) Prior to the cross-coupling, however, the energetically higher *s-cis* conformation should be adopted – this problem has been noted in the context of IMDA reactions.[154] On the other hand, amides have a lower energy barrier towards the *s-cis* conformation due to an $A^{1,3}$-strain between the nitrogen residue and the carbonyl oxygen. The use of a macrocyclic amide, although promising, is not appropriate for the overall synthetic plan. However, both aspects, the overall synthetic plan and the preference for the *s-cis* conformation can be combined when an *N*-benzyl hydroxamic ester is incorporated into the synthesis. (Scheme 122) *N*-Benzyl hydroxamic esters have already been successfully applied in IMDA reactions;[155] therein, the preferred *s-cis* conformation of the hydroxamic ester was proven both by calculations and the experimental outcome. Therefore, the intramolecular Stille-coupling of the vinyl iodide with the vinyl stannane should be facilitated – in the further synthetic sequence the hydroxamic ester poses no problem – after reduction to the cyclic aminal, the N-O bond can be cleaved with Zn/HOAc to give lactol **345**, or a related molecule.

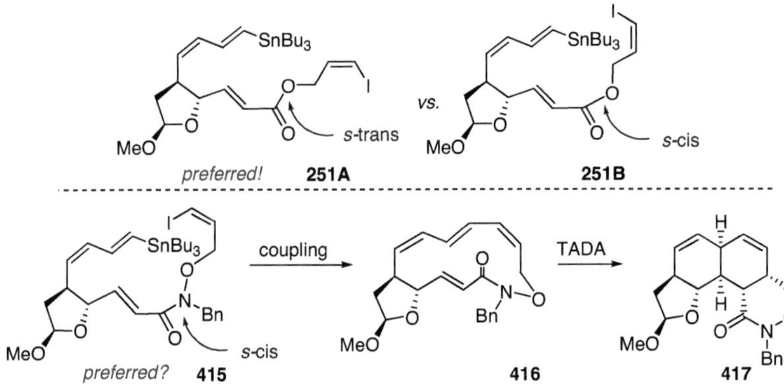

Scheme 122: Improved coupling using an *N*-benzyl hydroxamic ester

---

[154] a) Saito, A.; Ito, H.; Taguchi, T. *Org. Lett.* **2002**, *4*, 4619 – 4621; b) Saito, A.; Yanai, H.; Taguchi, T. *Tetrahedron Lett.* **2004**, *45*, 9439 – 9442; c) Saito, A.; Yanai, H.; Taguchi, T. *Tetrahedron* **2004**, *60*, 12239 – 12247.

[155] Ishikawa, T.; Senzaki, M.; Kadoya, R.; Morimoto, T.; Miyake, N.; Izawa, M.; Saito, S.; Kobayashi, H. *J. Am. Chem. Soc.* **2001**, *123*, 4607 – 4608.

9.4 Proposed Endgame

As nearly all problems in the synthesis of the main building blocks have been solved in this thesis, it seems appropriate to discuss the remaining steps necessary for the completion of the total synthesis of branimycin (**14**). (Scheme 123) Addition of the lithiated side-chain **411** to the octalin core fragment **336** should directly give the oxo-bridged octalin **418**. After protection of the newly formed secondary alcohol as the TBS-ether, the primary TBDPS-group will be selectively deprotected followed by a two-step oxidation to the acid. Deprotection of the PMB-ether with DDQ will then give the *seco*-acid **419**. The 9-membered macrolactone ring shall then be closed under the Yamaguchi-macrolactonization conditions. The macrolactone probably now allows the diastereofacially controlled introduction of C-2 methoxymethylen group in two steps. Global deprotection will then give branimycin (**14**).

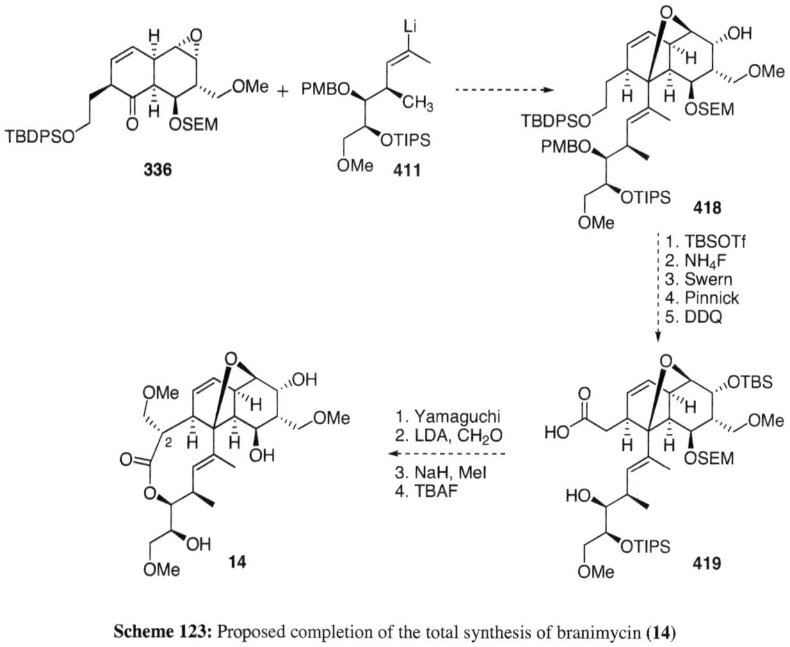

Scheme 123: Proposed completion of the total synthesis of branimycin (**14**)

EXPERIMENTAL SECTION

10 GENERAL

10.1 Solvent Purification

All reactions were carried out in oven or flame-dried glassware under an argon atmosphere, unless otherwise stated. Anhydrous tetrahydrofuran (THF) and diethyl ether (Et_2O) were freshly distilled from sodium/benzophenone under argon; anhydrous dichloromethane (DCM) and CH_3CN were freshly distilled from CaH_2 under argon. MeOH was distilled from magnesium turnings before use. Aceton was distilled from P_2O_5 and stored under 3Å molecular sieves. All other solvents were HPLC grade.

10.2 Reaction Control

Reactions were magnetically stirred and monitored by thin layer chromatography (TLC) with E. Merck silica gel 60-F_{254} plates. For the development of spots, ammonium molybdate/ceric sulfate (25 g ammonium molybdate, 0.4 g $CeSO_4$ in 250 mL H_2O), anisaldehyde (12 g anisaldehyde, 5 g H_2SO_4 in 300 mL EtOH) and permanganate (3 g $KMnO_4$ in 300 mL H_2O) dyes were used.

10.3 Column Chromatography

Flash column chromatography was performed with Merck silica gel (0.04-0.063 mm, 240-400 mesh), using as a rule of thumb 20-fold excess of silica gel based on crude product weight. As a routine, conditioning of the columns was performed by "wet-packing". When dry loading conditions were used, the crude product was dissolved in a small amount of DCM and silica gel was added (1.5 fold crude product weight). After careful removal of the solvent, the solid was loaded on an already conditioned column.

Yields refer to chromatographically and spectroscopically pure compounds, unless otherwise stated.

10.4 NMR-Spectroscopy

NMR spectra were recorded on either Bruker Avance DPX 250, DRX 400 or DRX 600 MHz spectrometer. Unless otherwise stated, all NMR spectra were measured in $CDCl_3$ solutions and referenced to the residual $CHCl_3$ signal (^1H, δ = 7.26; ^{13}C, δ = 77.0). All ^1H and ^{13}C shifts are given in ppm (s = singlet; d = doublet; t = triplet; q = quadruplet; m = multiplet; b = broad signal). Coupling constants J are given in Hz. In tin containing compounds, the ^{117}Sn and ^{119}Sn couplings were omitted. Assignments of proton resonances were confirmed, when possible, by correlated spectroscopy. CH-Multiplicity is based of routinely performed APT-experiments and is included where possible.

10.5 Other Spectroscopic Methods

Optical rotations were measured on a P 341 Perkin-Elmer polarimeter in a 10 cm cuvette at 20 °C with 589 nm wavelength.

Infrared spectra were acquired on a Perkin-Elmer 1600 FTIR-spectrometer; absorptions are given in wavenumber units [cm^{-1}]. The samples were measured as thin films on silicon plates.

Mass spectra were measured on a Micro mass, trio 200 Fisions Instruments. High resolution mass spectra (HRMS) were performed with a Finnigan MAT 8230 with a resolution of 10000.

10.6 X-Ray Analysis

Single crystal diffractions were collected on a Bruker X8APEX II CCD diffractometer. The structure was solved by direct methods and refined by full-matrix least-squares techniques. Non-hydrogen atoms were refined with anisotropic displacement parameters. H atoms were placed at calculated positions and refined as riding atoms in the subsequent least squares model refinements. Structure solution and refinement was performed with the SHELX program. (G.M. Sheldrick, *Program for crystal structure solution*, Universität Göttingen, **1997**; G.M. Sheldrick, *Program for crystal structure refinement*, Universität Göttingen, **1997**)

11 EXPERIMENTAL PROCEDURES

11.1 First Generation Approach

(*3aR,6R,6aS*)-6-(hydroxymethyl)-2,2-dimethyldihydrofuro[3,4-*d*][1,3]dioxol-4(3aH)-one[48]

Iodine (0.51 g, 2.0 mmol) was dissolved in acetone (80 mL). After addition of D-ribonolactone (**110**) (2.0 g, 13.3 mmol), the resulting solution was stirred at room temperature over night.

The reaction mixture was diluted with 100 mL of DCM, and then washed twice with 1M $Na_2S_2O_3$. The aqueous layer was extracted once with DCM. The combined organic layers were washed with brine, dried over $MgSO_4$, filtered and concentrated under reduced pressure. The crude residue was dissolved in hot EtOAc and, after cooling, precipitated by the addition of hexanes to yield pure **111** (2.18 g, 87%)

^{1}H NMR (250 MHz, CDCl$_3$): δ = 4.82 (d, *J* = 5.5 Hz, 1 H), 4.77 (d, *J* = 5.5 Hz, 1 H), 4.62 (dd, *J* = 1.9, 1.9 Hz, 1 H), 3.98 (ddd, *J* = 2.3, 5.3, 12.3 Hz, 2 H), 3.79 (ddd, *J* = 12.3, 5.7, 1.8 Hz, 1 H), 2.78 (t, *J* = 5.5 Hz, 1 H), 1.46 (s, 3 H), 1.37 (s, 3 H). *cf.* Ref. 48

R$_f$ (Hex/EtOAc = 1/2): 0.36

(3aS,4S,6aR)-2,2-dimethyl-6-oxotetrahydrofuro[3,4-d][1,3]dioxole-4-carbaldehyde[50]

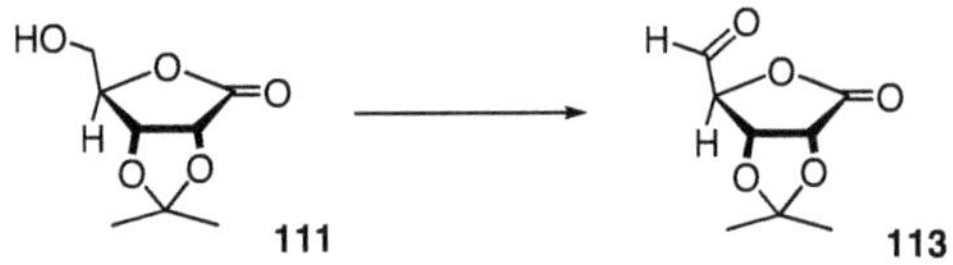

Dess-Martin periodinane (4.63 g 10.9 mmol) was dissolved in 40 mL of dichloromethane and **111** (1.87 g, 9.9 mmol) was added. After stirring at r.t. for 2 h, reaction control via TLC showed complete consumption of starting material.

400 mL of diethyl ether were added and resulting slurry was stirred for 10 min, after which it was filtered through a pad of celite. The solvents were then removed under reduced pressure. Purification by column chromatography (Hex:EtOAc = 1:1 to 1:2) yielded **113** as a colourless oil. (1.69 g, 91 %)

R_f (Hex/EtOAc = 1/2): 0.21

(E)-3-((3aR,4R,6aR)-2,2-Dimethyl-6-oxo-tetrahydro-furo[3,4-d][1,3]dioxol-4-yl)-propenal

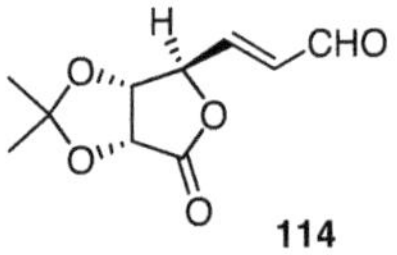

^{1}H NMR (250 MHz, CDCl$_3$): δ = 9.26 (d, *J* = 7.3 Hz, 1 H), 6.80 (dd, *J* = 15.9, 4.0 Hz, 1 H), 6.37 (ddd, *J* = 15.9, 7.2, 1.4 Hz, 1 H), 5.30 (d, *J* = 2.3 Hz, 1 H), 3.99 (dd, *J* = 12.1, 2.3 Hz, 1 H), 3.81 (d, *J* = 12.1, 1 H), 1.52 (s, 3 H), 1.45 (s, 3 H).

R_f (Hex/EtOAc = 2/1): 0.69

(3aR,6R,6aR)-6-(*(E)*-2-Iodo-vinyl)-2,2-dimethyl-dihydro-furo[3,4-*d*][1,3]dioxol-4-one

117

^{1}H NMR (250 MHz, CDCl$_3$): δ = 6.68 (dd, *J* = 14.7, 1.3 Hz, 1 H), 6.55 (dd, *J* = 14.7, 4.8 Hz, 1 H), 4.94 (d, *J* = 4.8 Hz, 1 H), 4.74 (d, *J* = 5.6 Hz, 1 H), 4.61 (d, *J* = 5.6 Hz, 1 H), 1.48 (s, 3 H), 1.37 (s, 3 H).

R$_f$ (Hex/EtOAc = 3/1): 0.57

1-Iodo 3-methoxy-1-propyne[53b]

126

3-Methoxy-1-propyne[156] was dissolved in 10 mL of THF. After cooling to -78 °C, *n*-BuLi in hexanes (3.15 mL, 7.89 mmol, 2.5 M) was added slowly. After stirring at − 78 °C for 1 h, a solution of iodine (2.0g, 7.89 mmol) in 5 mL of THF was added. After warming the reaction mixture to r.t., about 2/3 of the solvent was removed under reduced pressure. The residue was diluted with diethyl ether and then washed with 1M Na$_2$S$_2$O$_3$ solution. The aqueous layer was then extracted 4 times with diethyl ether. The combined organic layers were washed with brine, dried over MgSO$_4$, filtered and carefully concentrated under reduced pressure. Distillation under reduced pressure yielded pure **126** as a colourless liquid. (1.28 g, 93%)

^{1}H NMR (250 MHz, CDCl$_3$): δ = 4.12 (s, 2 H), 3.37 (s, 3 H). *cf*. Ref. 157

---

[156] Reppe, W. *et al*. *Liebigs Ann. Chem.* **1955**, *596*, 38 – 79.
[157] Barluenga, J.; González, J. M.; Rodríguez, M. A.; Campos, P. J.; Asensio, G. *Synthesis*, **1987**, 661 – 662.

1-Iodo 3-methoxy Z-propene

127: I–CH=CH–CH$_2$–OMe

Dipotassium azodicarboxylate was freshly prepared prior to the reaction.[53a] Azodicarboxylic acid diamide (1.16 g, 10 mmol) was added slowly over a 30 min period to 7.2 mL of 40% KOH at 0 °C. After stirring at 0 °C for 2 h, the bright yellow precipitate was filtered off and then washed with MeOH to yield the dipotassium azodicarboxylate. (1.40 g, 90%)

Compound **126** (1.28 g, 6.53 mmol) was dissolved in a mixture 9 mL of MeOH and 3.3 mL of pyridine. Dipotassium azodicarboxylate (1.26 g, 8.16 mmol) was added and the mixture cooled to 0 °C. Acetic acid (0.94 g, 0.9 mL, 15.67 mmol) was added *via* syringe pump over a 16 h period.

After the addition was finished, the reaction was carefully quenched by addition of 1M HCl. After confirmation that the pH was ~5, the mixture was extracted 4 times with diethyl ether. The combined organic layers were washed with brine, dried over MgSO$_4$, filtered and carefully concentrated under reduced pressure. The crude product was used without any further purification for the next step. (1.18 g, 92%)

^{1}H NMR (250 MHz, CDCl$_3$): δ = 6.40 (m, 2 H), 4.01 (m, 2 H), 3.36 (s, 3 H).

^{13}C NMR (63 MHz, CDCl$_3$): δ = 138.5 (CH), 83.4 (CH), 75.4 (CH$_2$), 58.7 (CH$_3$).

R$_f$ (Hex/EtOAc = 10/1): 0.33

((Z)-5-Methoxy-pent-3-en-1-ynyl)-trimethyl-silane

128: TMS–C≡C–CH=CH–CH$_2$–OMe

A 50 mL Schlenk flask was charged with 20 mL of freshly destilled NEt$_3$. Then PdCl$_2$(CH$_3$CN)$_2$ (0.18 g, 0.25 mmol) was added, followed by compound **127** (0.71 g, 3.58 mmol) dissolved in 2 mL of NEt$_3$. After stirring for 5 min, TMS-acetylen (0.65 g, 0.94 mL, 6.63 mmol) and CuI (0.23 g, 1.53 mmol) were added, upon which the heterogeneous solution turned from yellow to green-brown. After stirring at r.t. for 1.5 h, the solvent was removed under removed pressure.

The residue was redissolved in diethyl ether and washed twice with 1M HCl, followed by NaHCO$_3$, and brine. After drying over MgSO$_4$, the solvent was removed under reduced pressure. Purification by column chromatography (Hex:EtOAc = 20:1) yielded **128** as a colourless oil. (0.61 g, 93 %)

^{1}H NMR (250 MHz, CDCl$_3$): δ = 6.04 (td, *J* = 6.3, 11.1 Hz, 1 H) 5.64 (dt, *J* = 1.5, 11.1 Hz, 1 H), 4.19 (dd, J=1.5, 6.3 Hz, 2 H), 3.35 (s, 3 H), 0.19 (s, 9 H).

^{13}C NMR (63 MHz, CDCl$_3$): δ = 140.9 (CH), 112.0 (CH), 88.4 (C), 86.3 (C), 70.5 (CH$_2$), 58.5 (CH$_3$), -0.1 (CH$_3$).

R$_f$ (Hex/EtOAc = 10/1): 0.64

(*3aR,6R,6aR*)-2,2-Dimethyl-6-trityloxymethyl-dihydro-furo[3,4-*d*][1,3]dioxol-4-one[52]

Compound **111** (14.38 g, 76.4 mmol) and triethylamine (17.08 g, 23.4 mL, 168.1 mmol) were dissolved in 150 mL of DCM. A solution of trityl chlorid (23.4 g, 84.1 mmol) and DMAP (0.47 g, 3.82 mmol) in 100 mL of DCM was added over 30 min *via* a dropping funnel. After stirring at r.t. for 60 h, the reaction was quenched by addition of sat. NH$_4$Cl solution and then extracted 4 times with DCM. The combined organic layers were washed with brine, dried over MgSO$_4$, filtered and carefully concentrated under reduced pressure. The crude product was recrystallised from MeOH to give **131** as colourless crystals. (25.3 g, 77 %)

^{1}H NMR (250 MHz, CDCl$_3$): δ = 7.3 (m, 15 H), 4.97 (d, J = 5.7 Hz, 1 H), 4.58 (dd, J = 2.1, 2.3 Hz, 1 H), 4.44 (d, J = 5.7 Hz, 1 H), 3.72 (dd, J = 2.3, 10.7 Hz, 1 H), 3.10 (dd, J = 1.7, 10.7 Hz, 1 H), 1.47 (s, 3 H), 1.34 (s, 3 H).
^{13}C NMR (63 MHz, CDCl$_3$): δ = 174.7 (C), 143.3 (C), 128.9 (CH), 128.6 (CH), 127.8 (CH), 113.6 (C), 88.3 (C), 81.8 (CH), 79.0 (CH) 76.2 (CH), 63.3 (CH$_2$), 27.2 (CH$_3$), 26.0 (CH$_3$).

MS (ESI) m/z 551.2 (M+Na, 50)

R$_f$ (Hex/EtOAc = 3/1): 0.55

(3aR,6R,6aR)-2,2-Dimethyl-4-trimethylsilanylethynyl-6-trityloxymethyl-tetrahydro-furo[3,4-*d*][1,3]dioxol-4-ol

A 100 mL Schlenk flask was charged with 50 mL of THF and TMS-acetylen (0.92 g, 1.33 mL, 9.41 mmol). After cooling to – 78°C, *n*-BuLi in hexanes (3.8 mL, 9.41 mmol, 2.5M) was added slowly. After stirring at – 78°C for 30 min, the solution was cannulated into a second flask, containing compound **123** (1.0g, 4.18 mmol) dissolved in 50 mL of THF. After the addition was completed, the reaction was quenched by the addition of sat. NH$_4$Cl at – 78°C. After warming to r.t., the reaction mixture was concentrated under reduced pressure. The resulting mixture was then extracted 4 times with diethyl ether. The combined organic layers were washed with brine, dried over MgSO$_4$, filtered and concentrated under reduced pressure. Purification by column chromatography (Hex:EtOAc = 10:1 to 1:1) yielded **131** as a colourless oil. (1.63 g, 74%)

^{1}H NMR (400 MHz, CDCl$_3$) : (mixture of diastereomic ketals) δ = 7.40-7.15 (m, 15 H), 4.70 (d, J = 6.6 Hz, 0.5 H), 4.68 (dd, J = 5.9, 1.1 Hz, 0.5 H), 4.59 (dd, J = 6.6, 2.2 Hz, 0.5 H), 4.52 (d, J = 5.9 Hz, 0.5 H), 4.33 (dd, J = 4.4, 4.3 Hz, 0.5 H), 4.22 (ddd, J = 6.0, 4.0, 2.2 Hz, 0.5 H), 4.15 (s, 0.5 H), 3.84 (s, 0.5 H), 3.33 (dd, J = 9.7, 6.0 Hz, 0.5 H), 3.28 (dd, J = 8.6,4.7 Hz, 0.5 H), 3.25 (dd, J = 10.2, 4.9 Hz, 0.5 H), 3.17 (dd, J = 10.2, 4.1, Hz, 0.5 H), 1.52 (s, 1.5 H), 1.47 1.52 (s, 1.5 H), 1.30 (s, 3 H).

^{13}C NMR (100 MHz, CDCl$_3$): (mixture of diastereomic ketals) δ = 144.2 (C), 143.5 (C), 129.2 (CH), 129.1 (CH), 128.5 (CH), 128.3 (CH), 127.8 (CH), 127.5 (CH), 115.4 (C), 113.7 (C), 103.7 (C), 102.1 (C), 100.6 (C), 96.8 (C), 91.8 (C), 89.6 (C), 88.4 (C), 88.3 (CH), 87.4 (CH), 85.7 (CH), 82.9 (CH), 82.6 (CH), 82.1 (CH), 65.3 (CH$_2$), 63.9 (CH$_2$), 27.1 (CH$_3$), 26.6 (CH$_3$), 26.2 (CH$_3$), 25.4 (CH$_3$), 0.1 (CH$_3$), 0.0 (CH$_3$).

R$_f$ (Hex/EtOAc = 3/1): 0.41

((*1R,2R,6R,7R*)-4,4-Dimethyl-3,5,8,10-tetraoxa-tricyclo[5.2.1.0^{2,6}]dec-7-ylethynyl)-trimethyl-silane

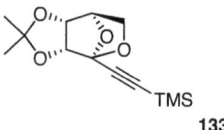

133

^{1}H NMR (250 MHz, CDCl$_3$): δ = 4.63 (d, J = 3.6 Hz, 1 H), 4.38 (d, J = 5.5 Hz, 1 H), 4.27 (d, J = 5.5 Hz, 1 H), 3.56 (dd, J = 7.1, 3.6 Hz, 1 H), 3.38 (d, J = 7.1 Hz, 1 H), 1.47 (s, 3 H), 1.29 (s, 3 H).

^{13}C NMR (63 MHz, CDCl$_3$): δ = 113.2 (C), 100.1 (C), 94.8 (C), 84.2 (CH), 80.5 (CH), 78.5 (CH), 64.5 (CH$_2$), 26.5 (CH$_3$), 26.0 (CH$_3$), -0.1 (CH$_3$).

MS (EI) m/z 253 (M⁺, 21), 210 (26), 195 (13), 151 (8), 125 (100).

HRMS(EI) calcd. for $C_{13}H_{20}O_4Si$: 268.1131, Found: 268.1126.

R_f (Hex/EtOAc = 3/1): 0.50

Acetic acid (R)-1-{(R)-5-[1-acetoxy-3-trimethylsilanyl-prop-2-yn-(E)-ylidene]-2,2-dimethyl-[1,3]dioxolan-4-yl}-2-trityloxy-ethyl ester

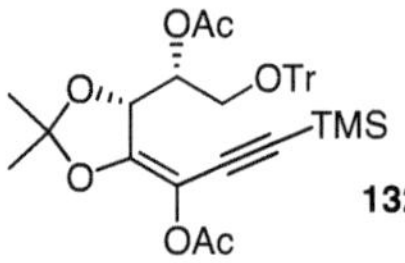

¹H NMR (400 MHz, CDCl₃): δ = 7.19 (d, J = 7.5 Hz, 6 H), 7.08 (t, J = 7.50 Hz, 6 H) 7.01 (t, J = 7.5 Hz, 3 H), 5.35 (ddd, J = 7.6, 4.0, 1.7 Hz, 1 H), 4.66 (d, J = 1.7 Hz, 1 H), 3.09 (dd, J = 10.1, 7.6 Hz, 1 H), 2.96 (dd, J=10.1, 4.0 Hz, 1 H).

¹³C NMR (100MHz, CDCl₃): δ = 170.5 (C), 168.9 (C), 152.0 (C), 143.8 (C), 128.8 (CH), 128.3 (CH), 128.0 (CH), 127.1 (CH), 114.9 (C), 111.0 (C), 102.2 (C), 95.5 (C), 87.2 (C), 78.6 (CH), 71.9 (CH), 61.9 (CH₂), 25.6 (CH₃), 25.4 (CH₃), 21.2 (CH₃), 20.8 (CH₃), 0.0 (CH₃).

R_f (Hex/EtOAc = 3/1): 0.30

1-((4*S*,5*R*)-5-((*R*)-1-hydroxy-2-(trityloxy)ethyl)-2,2-dimethyl-1,3-dioxolan-4-yl)-3-(trimethylsilyl)prop-2-yn-1-ol

Compound **131** (1.03 g, 1.95 mmol) and NH$_4$Cl (0.13 g, 2.34 mmol) were dissolved in 30 mL of MeOH and NaBH$_4$ (0.11 g, 2.92 mmol) was added. After stirring for 20 min, TLC showed some starting material present, so a second portion of NaBH$_4$ (0.02 g, 0.59 mmol) was added.

10 minutes later, the reaction was quenched by the addition of solid NH$_4$Cl. The reaction mixture was concentrated under reduced pressure, then the residue was redissolved in H$_2$O and then extracted 4 times with EtOAc. The combined organic layers were washed with brine, dried over MgSO$_4$, filtered and carefully concentrated under reduced pressure. Purification by column chromatography (Hex:EtOAc = 4:1) yielded **136** as a colourless oil. (0.96 g, 93%) diasteromeric ratio 85 : 15.

^{1}H NMR (250 MHz, CDCl$_3$): δ = 7.45 (m, 6 H), 7.34-7.21 (m, 9 H), 4.62 (dd, *J* = 8.9, 4.6 Hz, 1 H), 4.50 (m, 1 H), 4.32 (dd, *J* = 6.2, 4.3 Hz, 1 H), 4.23 (dd, *J* = 9.4, 6.2 Hz, 1 H), 3.67 (bd, *J* = 8.7 Hz, 1 H), 3.45 (dd, *J* = 9.6, 2.7 Hz, 1 H), 3.27 (dd, *J* = 9.6, 2.7 Hz, 1 H), 2.90 (d, *J* = 4.80, 1 H), 1.43 (s, 3 H), 1.33 (s, 3 H), 0.19 (s, 9 H).

R$_f$ (Hex/EtOAc = 3/1): 0.29

11.2 Second Generation Approach

L-Gulono-1,4-lactone

A solution of L-ascorbic acid (60.0 g, 344.5 mmol) in water (440 mL) was hydrogenated over Pd/C (4.8 g, 10%) in a Parr hydrogenation apparatus at 50°C and 50 psi hydrogen pressure. Reaction control by ^{13}C-NMR (ca. 200µl sample in d^6-DMSO) showed complete consumption of the starting material after 24h. The catalyst was removed by filtration through a pad of celite and the water was evaporated in vacuo to afford 60.0 g of a white solid. The white solid was suspended in 90 mL ethanol (p.a.) and was heated to reflux for 30 min and subsequently stored at –30 °C overnight. Filtration yielded **144** as colourless crystals (48.7 g, 88%), which were used without any further purification.

^{13}C NMR (100 MHz, DMSO): δ = 81.6 (CH), 71.6 (CH), 71.0 (CH), 70.3 (CH), 62.8 (CH$_2$).

5,6-*O*-Isopropylidene- L -gulono-1,4-lactone[57]

L-Gulono-1,4-lactone (**144**) (26.0g, 146.0mmol) was dissolved in anhydrous DMF (400 ml). The reaction mixture was cooled in an ice bath and treated with concentrated sulfuric acid (2.14 g, 1.14 mL, 21.8 mmol). The pH of ~ 3 was checked with pH-paper

(100 µl reaction mixture + 250 µl water) to decrease to a value of 3 (additional acid may be necessary). 2-Methoxypropene (12.7 g, 176.1 mmol, 1.2 eq) were added and the cooling bath was removed. The reaction mixture was stirred at room temperature until reaction control by ^{13}C-NMR (100 µl reaction mixture in DMSO) showed complete consumption of the starting material (approx. 8h). Sodium carbonate (2.31 g, 21.8 mmol) and water (19.0 ml) were added and stirring was continued for 1h. The solvent was removed after filtration through a pad of celite under reduced pressure to yield 34 g of a red solid, which was repeatedly recrystallised from ethyl acetate to yield a colorless powder. (25 g, 76%)

^{13}C NMR (100 MHz, DMSO): 176.8 (C), 109.8 (C), 82.1 (CH), 75.9 (CH), 71.1 (CH), 70.0 (CH), 65.2 (CH$_2$), 27.4 (CH$_3$), 26.1 (CH$_3$).

Ethyl (Z)-3-[(R)-2,2-Dimethyl-1,3-dioxolan-4-yl]-acrylate[57]

A suspension of NaIO$_4$ (53.4 g, 250.0 mmol) in water (126 mL) was cooled in an ice bath and treated with 3N sodium hydroxide solution (83.3 mL, 250.0 mmol). After the pH was adjusted to 5.5 (pH electrode, use 15% sodium carbonate solution and 32% hydrochloric acid), the ice bath was replaced by a water bath. Finely powdered 5,6-O-Isopropylidene-L-gulono-1,4-lactone (**145**) (125.0 mmol) was added at once (slightly exothermic reaction). The pH of the suspension was continously monitored and adjusted (15% sodium carbonate solution/1N hydrochloric acid) over 60 minutes. The reaction mixture was saturated with sodium chloride, filtered and the residue was then washed twice with brine. The aqueous filtrate was treated with a solution of (triphenyl phosphoranyliden)-acetic acid ethyl ester **138** (44.0 g, 126.3 mmol) in dichloromethane (400 mL). After vigorous stirring for 10 hours at room temperature the phases were separated and the aqueous phase was extracted with DCM. The combined organic layers were subsequentely washed with sat. NaHCO$_3$ solution, brine, dried over MgSO$_4$ and the solvent removed under reduced pressure. Purification by column chromatography (hexane/ ethyl acetate 7:1 to 3:1) yielded **147** and **148** in a 3: 1 ratio. (**147**: 17.0 g, 68%, **148**: 5.69 g, 23%)

Z-Isomer (**147**)
^{1}H NMR (250 MHz, CDCl$_3$): δ =6.12 (dd, J = 11.6, 6.5 Hz, 1 H), 5.59 (dd, J = 11.6, 1.7 Hz, 1 H), 5.23 (1 H, m), 4.11 (dd, J = 8.2, 7.0 Hz, 1 H), 3.92 (q, J = 7.1 Hz, 2 H), 3.34 (dd, J = 8.2, 6.8 Hz, 1 H), 1.18 (3 H, s), 1.12 (3 H, s), 1.03 (t, J = 7.1 Hz, 3 H).

^{13}C NMR (63 MHz, CDCl$_3$, ppm): δ = 165.5 (C), 149.7 (CH), 120.7 (CH), 109.6 (C), 73.7 (CH), 69.5 (CH), 60.3 (CH$_2$), 26.6 (CH$_3$), 15.5 (CH$_3$), 14.2 (CH$_3$).

E-Isomer (**148**)
^{1}H NMR (250 MHz, CDCl$_3$): δ = 6.79 (dd, J = 15.6, 5.6 Hz, 1 H), 6.00 (d, J = 15.6 Hz, 1 H), 4.57 (m, 1 H), 4.10 (m, 3 H), 1.35 (s, 3 H), 1.31 (s, 3 H), 1.20 (t, J = 7.1 Hz, 1 H).

^{13}C NMR (63 MHz, CDCl$_3$): δ = 166.3 (C), 144.9 (CH), 122.7 (CH), 110.4 (C), 75.2 (CH$_2$), 69.1 (CH), 60.8 (CH), 26.7 (CH$_3$), 26.0 (CH$_3$), 14.5 (CH$_3$).

Enantiomeric excess: > 99.5
HPLC: Chiralcel OD-H, 5°C, hexane/*iso*-propanol 98:2, 0.5 ml/min

Z-olefin, *R*-enantiomer	8.1 min	Z-olefin, *S*-enantiomer	14.1 min
E-olefin, *R*-enantiomer	12.5 min	E-olefin, *S*-enantiomer	20.1 min

(R)-5-Hydroxymethyl-5H-furan-2-one

To a solution of **147** (4 g, 20 mmol) in ethanol (25 mL) was added 32% hydrochloric acid (1 mL, 10 mmol). The reaction mixture was stirred at room temperature until TLC showed complete consumption of the starting material (approx. 2.5h). The reaction mixture was cooled to 0°C and dropwise treated with triethylamine (1.39 ml, 10 mmol). The pH-value of the final solution was checked to be in the neutral range (pH-paper, around 6). To the reaction mixture was added toluene (250 mL). After removal of the solvents the crude product was purified by column chromatography (hexane/ethyl acetate 1:1 to 1:3) to yield **151** as a colourless, solidifying oil. (2.14 g, 94%).

^{1}H NMR (250 MHz, CDCl$_3$): δ = 7.49 (dd, J = 5.7, 1.5 Hz, 1 H), 6.14 (dd, J = 5.7, 2.0 Hz, 1 H), 5.13 (m, 1 H), 3.94 (dd, J = 12.2, 3.5 Hz, 1 H), 3.75 (dd, J = 12.2, 4.8 Hz, 1 H), 3.50 (s, OH).

^{13}C NMR (63 MHz, CDCl$_3$): δ = 173.9 (C), 154.4 (CH), 123.0 (CH), 84.6 (CH), 62.4 (CH$_2$).

R$_f$ (Hex/EtOAc = 1/3) 0.18

(S)-5-Hydroxymethyl-4-phenylsulfanyl-dihydro-furan-2-one[59]

Compound **148** (8.69 g, 43.4 mmol) was dissolved in benzene (80 mL). Thiopenol (8.9 mL, 86.8 mmol) and ethyldiisopropylamine (15.2 ml, 86.8 mmol) were added and the solution was stirred until TLC showed complete consumption of the starting material. After removal of the solvent in vacuo, the residue is purified by column chromatography (hexane/ethyl acetate 9:1 → 5:1) to yield the desired product in quantitative yield as 3:2 mixture of diastereomers. This mixture was employed in the next step without separation.

Anti-adduct:

^{1}H NMR (250 MHz, CDCl$_3$): δ = 7.50 (m, 2 H), 7.30 (m, 3 H), 4.15 (m, 3 H), 4.05 (dd, *J* = 9.0, 6.0 Hz, 1 H), 3.81 (dd, *J* = 8.0, 5.2 Hz, 1 H), 3.44 (ddd, *J* = 16.0, 9.0 Hz, 1 H), 2.90 (dd, *J* = 16.0, 5.2 Hz, 1 H), 2.54 (dd, *J* = 16.0, 9.0 Hz, 1 H), 1.39 (s, 3 H), 1.31 (s, 3 H), 1.26 (t, *J* = 7.1 Hz, 3 H).

Syn-adduct:

^{1}H NMR (250 MHz, CDCl$_3$): δ = 7.50 (m, 2 H), 7.30 (m, 3 H), 4.36 (dt, *J* = 6.6, 4.3 Hz, 1 H), 4.60 (q, *J* = 7.4 Hz, 2 H), 4.07 (dd, *J* = 8.6, 6.6 Hz, 1 H), 3.87 (dd, 8.6, 6.6 Hz, 1 H), 3.79 (ddd, *J* = 8.7, 5.7, 4.3 Hz, 1 H), 2.84 (dd, *J* = 16.2, 5.7 Hz, 1 H), 2.53 (dd, *J* = 16.2, 8.7 Hz, 1 H), 1.39 (s, 3 H), 1.31 (s, 3 H), 1.26 (t, *J* = 7.1 Hz, 3 H).

The crude product was dissolved in ethanol (6.5mL). Concentrated hydrochloric acid (32%, 700 μl) were added an the reaction mixture was stirred until TLC showed completion of the reaction (ca. 24h). After most of the ethanol had been evaporated, the residue was diluted with water and then extracted 4 times with DCM. The combined organic layers were washed with brine, dried over MgSO$_4$, filtered and concentrated under reduced pressure. The crude product **156a** was directly used without any further purification. (8.86 g, 91%)

R$_f$ (Hex/EtOAc = 3/1): 0.11

(R)-5-Hydroxymethyl-5H-furan-2-one (Method B)

A solution of sodium periodate (200 mg, 0.95 mmol) in water (5ml) was added dropwise to a pre-cooled (0°C) solution of **156a** (200 mg, 0.9 mmol) in methanol (5 ml). After the addition was completed, the cooling bath was removed and stirring continued until TLC (hexane/ethyl acetate 1:3) showed complete consumprtion of the starting material. The precipitate was filtered and the filtrate extracted with dichloromethane. The organic phase was washed with brine and dried over magnesium sulfate to yield the desired product in quantitative yield. (214mg, 94%)

R_f (Hex/EtOAc = 1/3): 0.42

A suspension of **157** (1.46 g, 6.1 mmol) and calcium carbonate (2.25 g, 22.5 mmol) in toluene (15 ml) was heated to reflux until TLC showed complete consumption of the starting material. After filtration through celite and removal of the solvent under reduced pressure, purification by column chromatography (Hex:EtOAc = 1:1 to 1:3) yielded **151** as a colourless, solidifying oil. (0.56 g, 80%).

Spectroscopic analysis was identical to that of **151** obtained from **147**.

(R)-5-(tert.-Butyl-diphenyl-silanyloxymethyl)-5H-furan-2-one

A solution of **151** (200 mg, 1.75 mmol) and imidazole (133 mg, 1.95 mmol) in DMF (0.6 mL) was cooled to 0 °C and treated dropwise with *tert*-butyl diphenyl silylchloride (0.5 mL, 1.95 mmol). After 15 minutes the reaction mixture was warmed to r.t. The stirring was continued until TLC showed complete consumption of the starting material (ca. 30 min).

The reaction was quenched by addition of saturated $NaHCO_3$ solution (4 mL) and EtOAc. The organic phase was separated and subsequently washed with water and brine. The organic solvents were evaporated after addition of toluene (50 mL) and the remaining oil was purified by column chromatography using Hex:EtOAc = 5:1 to 2:1 as eluent to yield **152** (580 mg, 94 %) as colourless solid.

HPLC: Chiralcel OD-H, 5 °C, R-enantiomer: 18.2 min, S-enantiomer: 16.3 min (Hex/iso-propanol 95:5, 0.8 ml/min): ee: > 99.5 %.

^{1}H NMR (400 MHz, $CDCl_3$,): δ =7.64 (m, 4 H), 7.42 (m, 7 H), 6.18 (dd, *J* = 5.8, 2.0 Hz, 1 H), 5.07 (m, 1 H), 3.92 (dd, *J* = 10.9, 4.5 Hz, 1 H), 3.88 (dd, *J* = 10.9, 4.9 Hz, 1 H), 1.04 (s, 9 H).

^{13}C NMR (100 MHz, $CDCl_3$): δ = 173.3 (C), 154.4 (CH), 136.0 (CH), 135.9 (CH), 133.2 (C), 132.9 (C), 130.4 (CH), 128.3 (CH), 123.1 (CH), 83.6 (CH), 63.8 (CH_2), 27.1 (CH_3), 19.6 (C).

IR [cm^{-1}]: 3094, 3070, 2955, 2933, 2858, 1771, 1748, 1604, 1588, 1485, 1474, 1465, 1427, 1378, 1360, 1330, 1243, 1162, 1129, 1099, 1062, 1013, 951, 917, 883, 836, 826.

HRMS(EI) calcd. for $C_{17}H_{15}O_3Si$: 295.0790, Found: 295.0797.

(4S,5R)-5-(tert.-Butyl-diphenyl-silanyloxymethyl)-4-vinyl-dihydrofuran-2-one

A 250ml Schlenk flask was charged with CuCl (103 mg, 1.04 mmol) and LiCl (94 mg, 2.22 mmol) and stirred under vacuum for 2 h. Then, 130 mL of THF were added to give a yellow-green solution. The solution was cooled to -78° C, vinylmagnesium chloride in THF (24.3 mL, 29.64 mmol, 1.22 M) was added and the solution was stirred for 10 min, after which a precooled solution of **152** (5.22 g, 14.82 mmol) in 40 mL of THF was added via cannula. After stirring for 40 min, the cooling bath was removed, the solution warmed to -50 °C and then quenched with 30 mL of 1M hydrochloric acid. After the reaction mixture had reached room temperature, it was diluted with ethyl acetate and brine containing 5% ammonia (25 wt% aqueous solution). The precipitate was removed by filtering through a pad of celite. The organic phase was washed with brine containing 5% ammonia (25 wt% aqueous solution), until the aqueous layer stayed colourless. After another wash with brine, the organic layer was dried over magnesium sulfate, filtered and the solvent removed under reduced pressure to yield **158** as a light-yellow oil (6.05 g), which was employed without further purification for the next step.

^{1}H NMR (250 MHz, CDCl$_3$): δ = 6.67 (m, 4 H), 7.42 (m, 6 H), 5.76 (ddd, J = 17.6, 9.6, 8.0 Hz, 1 H), 5.13 (d, J = 17.6 Hz, 1 H), 5.11 (d, J = 9.6 Hz, 1 H), 4.25 (ddd, J = 6.8, 3.5, 2.7 Hz, 1 H), 3.93 (dd, J = 11.6, 2.7 Hz, 1 H), 3.73 (dd, J = 11.6, 3.5 Hz, 1 H), 3.20 (dddd, J = 8.9, 8.4, 8.0, 6.8 Hz, 2.82 (dd, J = 17.6, 8.9 Hz, 1 H), 2.44 (dd, J = 17.6, 8.4 Hz, 1 H).

^{13}C NMR (63 MHz, CDCl$_3$): δ = 175.9 (C), 136.4 (CH), 136.0 (CH), 135.5 (CH), 132.9 (C), 132.5 (C), 129.9 (CH), 127.8 (CH), 117.3 (CH$_2$), 104.5 (CH), 84.6 (CH$_2$), 63.3 (CH), 40.7 (CH), 35.0 (CH$_2$), 26.7 (CH$_3$), 19.2 (C).

2,2-Dimethyl-propionic acid (R)-5-oxo-2,5-dihydro-furan-2-ylmethyl ester[158]

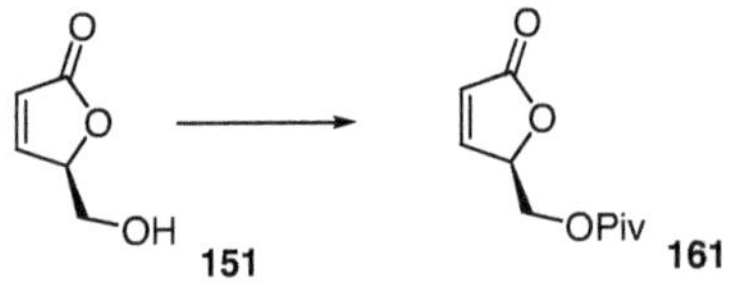

Compound **151** (400 mg, 3.51 mmol) was dissolved in 5 mL of DCM, cooled to -10 °C and pyridine (291 mg, 0.3 mL, 3.68 mmol) and PivCl (444 mg, 0.45 mL, 3.68 mmol) were added. The reaction mixture stirred at r.t. for 24h.

After dilution with DCM, the mixture was washed twice with water. The combined organic layers were washed with brine, dried over $MgSO_4$, filtered and concentrated under reduced pressure. Purification by column chromatography (Hex:EtOAc = 2:1) yielded **161** as a colourless oil. (660 mg, 95 %).

^{1}H NMR (250 MHz, $CDCl_3$): δ = 7.42 (dd, J = 5.7, 1.6 Hz, 1 H), 6.19 (dd, J = 5.7, 2.1 Hz, 1 H), 5.22 (ddt, J = 4.0, 2.2, 1.6 Hz, 1 H), 4.37 (d, J = 3.6 Hz, 2 H), 1.17 (s, 9 H).

^{13}C NMR (63 MHz, $CDCl_3$): δ = 176.2 (C), 152.3 (CH), 123.2 (CH), 81.0 (CH), 62.0 (CH_2), 27.0 (CH_3).

R_f (Hex/EtOAc = 3/1): 0.18

---

[158] Ondruš, V.; Orsága, M.; Fišera, L.; Prónayová, N. *Tetrahedron* **1999**, *55*, 10425 – 10436.

(3S,4R,5R)-3-(2,2-Dimethyl-propionyl)-5-hydroxymethyl-4-vinyl-dihydro-furan-2-one

A 50ml Schlenk flask was charged with CuCl (34 mg, 0.34 mmol) and LiCl (36 mg, 0.85 mmol) and stirred under vacuum for 2 h. Then, 7 mL of THF were added to give a yellow-green solution. The solution was cooled to -78 °C, vinylmagnesium chloride in THF (3.0 mL, 3.73 mmol, 1.22 M) was added and the solution was stirred for 10 min, after which a solution of **161** (366 mg, 1.70 mmol) in 7 mL of THF was added. After stirring for 1.5 h, the mixture was warmed to -50°C and kept at this temperature for another hour. After warming to 0 °C, the reaction was quenched with 3.5 mL of 1M hydrochloric acid. After the reaction mixture had reached room temperature, it was diluted with ethyl acetate and brine containing 5% ammonia (25 wt% aqueous solution). The precipitate was removed by filtering through a pad of celite. The organic phase was washed with brine containing 5% ammonia (25 wt% aqueous solution), until the aqueous layer stayed colourless. After another wash with brine, the organic layer was dried over magnesium sulfate, filtered and the solvent removed under reduced pressure. Purification by column chromatography (Hex:EtOAc = 4:1 to 2:1) yielded **162** as a colourless oil. (2.57 g, 67%)

^{1}H NMR (400 MHz, CDCl$_3$): δ = 5.63 (ddd, J = 16.7, 10.0, 8.9 Hz, 1 H), 5.16 (d, J = 16.7 Hz, 1 H), 5.15 (d, J = 10.0 Hz, 1 H), 4.24 (ddd, J = 9.5, 5.3, 2.5 Hz, 1 H), 4.05 (d, J = 10.6 Hz, 1 H), 3.87 (ddd, J = 12.7, 6.5, 2.6 Hz, 1 H), 3.69 (ddd, J = 12.7, 6.7, 5.7 Hz, 1 H), 3.43 (q, J = 9.7 Hz, 1 H), 1.95 (s, OH), 1.12 (s, 9 H).

^{13}C NMR (100 MHz, CDCl$_3$): δ = 197.6 (C), 172.1 (C), 134.0 (CH), 121.0 (CH$_2$), 84.4 (CH), 62.4 (CH$_2$), 54.0 (CH), 47.2 (CH), 45.1 (C), 25.8 (CH$_3$).

R$_f$ (Hex/EtOAc = 1/1): 0.33

(R)-3-Phenylsulfanyl-5-trityloxymethyl-dihydro-furan-2-one

At 0 °C, to a solution of diisopropylamine (2.01 g, 2.79 mL, 19.9 mmol) in 31 mL of THF was added n-BuLi in hexanes (12.26 mL, 19.6 mmol, 1.6M). After stirring for 30 minutes, a solution of **170c** (1.60 g, 9.49 mmol) in 5 mL of THF was slowly added. After stirring at 0 °C for 30 min, the solution was warmed to r.t. and stirring continued for another 30 min. Then, it was cooled to – 78 °C and a solution of **171b** (2 g, 6.33 mmol) in 31 mL THF was added. The reaction was stirred at -78 °C for one hour and then allowed to warm to r.t. overnight.

The reaction was diluted with diethyl ether and then poured onto 1N NaOH. The organic layer was washed another 2 times with 1M NaOH. The combined aqueous layers were carefully acidified to pH ~ 5 and then extracted 3 times with DCM. The combined organic layers (including the initial ethereal phase) were washed with brine, dried over $MgSO_4$, filtered and concentrated under reduced pressure.

The residue was then suspended in toluene and heated to reflux with a Dean-Stark trap for 1 hour. After removal of the solvents, purification by column chromatography (Hex:EtOAc = 7:1 to 4:1) yielded **178** as a colourless oil. (2.33 g, 79 %).

Major epimer:
^{1}H NMR (250 MHz, CDCl$_3$): δ = 7.40-7.15 (m, 20 H), 4.52 (ddd, J = 11.4, 5.0, 3.3 Hz, 1 H), 4.16 (dd, J = 9.3, 7.0 Hz, 1 H), 3.50 (dd, J = 10.5, 3.2 Hz, 1 H), 3.10 (dd, J = 10.5, 3.6 Hz, 1 H), 2.51 (ddd, 14.6 Hz, 8.9, 4.2 Hz, 1 H), 2.28 (m, 1 H).

R$_f$ (Hex/EtOAc = 3/1): 0.7 (major epimer); 0.58 (minor epimer)

(R)-3-Hydroxymethyl-3-phenylsulfanyl-5-trityloxymethyl-dihydro-furan-2-one

At 0 °C, to a solution of diisopropylamine (51 mg, 71 µl, 0.51 mmol) in 1 mL of THF was added *n*-BuLi in hexanes (0.32 mL, 0.51 mmol, 1.6M). After stirring at 0 °C for 30 min, the solution was cooled to -78 °C and a solution of **178** (180 mg, 0.39 mmol) in 2 mL of THF was added slowly. After stirring at -50 °C for one hour, formaldehyde gas (produced by pyrolysis of *para*-formaldehyde (56mg, 1.87 mmol) in an Argon stream) was introduced *via* a teflon cannula. After stirring at -50 °C for another hour, the reaction was warmed to r.t. overnight.

The reaction was quenched by the addition of sat. NH$_4$Cl solution and then extracted 4 times with diethyl ether. The combined organic layers were washed with brine, dried over MgSO$_4$, filtered and concentrated under reduced pressure. Purification by column chromatography (Hex:EtOAc = 7:1 to 4:1) yielded **179** as a colourless oil. (2:3 mixture of epimers, 119 mg, 62%)

Syn-epimer:
^{1}H NMR (400 MHz, CDCl$_3$): δ = 7.50-7.05 (m, 20 H), 4.65 (dddd, *J* = 10.1, 6.3, 4.3, 3.1 Hz, 1 H), 3.83 (dd, *J* = 11.4, 4.2 Hz, 1 H), 3.65 (dd, *J* = 11.4, 4.3 Hz, 1 H), 3.27 (dd, *J* = 10.5, 3.6 Hz, 1 H), 3.15 (dd, *J* = 10.5, 5.1 Hz, 1 H), 2.60 (dd, *J* = 13.7, 10.2 Hz, 1 H), 1.99 (dd, *J* = 13.7, 5.9 Hz, 1 H).

^{13}C NMR (100 MHz, CDCl$_3$): δ = 174.8 (C), 143.9 (C), 137.7 (CH), 130.7 (CH), 129.5 (CH), 129.0 (CH), 128.6 (C), 128.3 (CH), 127.6 (CH), 87.2 (C), 76.9 (CH), 64.6 (CH$_2$), 56.3 (CH$_2$), 33.8 (CH$_2$).

Anti-epimer:

¹H NMR (400 MHz, CDCl₃): δ = 7.40-7.15 (m, 20 H), 4.57 (dddd, *J* = 9.3, 6.1, 5.0, 4.1 Hz, 1 H), 3.81 (dd, *J* = 11.2, 4.9 Hz, 1 H), 3.62 (dd, *J* = 11.2, 3.5 Hz, 1 H), 3.31 (dd, *J* = 10.1, 7.0 Hz, 1 H), 3.06 (dd, *J* = 10.1, 4.7 Hz, 1 H), 2.60 (dd, *J* = 14.1, 8.5 Hz, 1 H), 2.00 (dd, *J* = 14.1, 5.3 Hz, 1 H).

¹³C NMR (100 MHz, CDCl₃): δ = 176.1 (C), 143.9 (C), 137.4 (CH), 130.4 (CH), 129.4 (CH), 129.2 (C), 129.1 (CH), 128.3 (CH), 127.5 (CH), 87.3 (C), 76.9 (CH), 65.9 (CH₂), 55.4 (CH₂), 33.5 (CH₂).

R$_f$ (Hex/EtOAc = 3/1): 0.36 (*syn*-adduct), 0.25 (*anti*-adduct)

(*R*)-3-Hydroxymethyl-5-trityloxymethyl-(5*H*)-furan-2-one

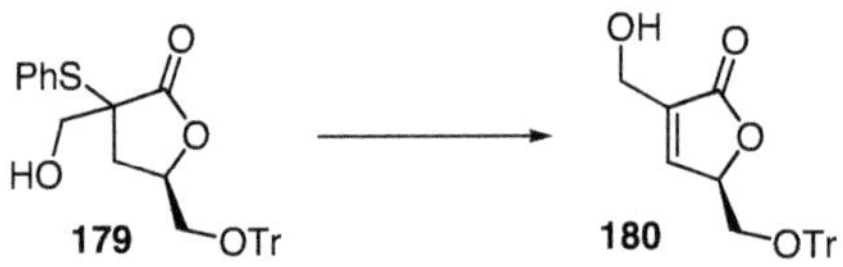

Compound **179** (68 mg, 0.14 mmol) was dissolved in 4 mL of DCM and cooled to 0 °C. *m*CPBA (27 mg, 0.16 mmol, 77%) was added and the reaction was stirred for 2 hours. The reaction was quenched by the addition of 2 mL of 1 M Na₂S₂O₃ and then extracted 4 times with DCM. The combined organic layers were washed with brine, dried over MgSO₄, filtered and concentrated under reduced pressure.

The crude product was dissolved in 5 of mL toluene, CaCO₃ (41 mg, 0.41 mmol) was added and the mixture heated to reflux for 2.5 h. After the mixture was cooled to r.t., it was filtered through a pad of celite and the solvent was removed under reduced pressure. Purification by column chromatography (Hex:EtOAc = 10:1) yielded **180** as a colourless oil. (26 mg, 48 %)

¹H NMR (250 MHz, CDCl₃): δ = 7.50-7.18 (m, 15 H), 5.10 (m, 2 H), 4.47 (s, 2 H), 3.41 (d, 5.2 Hz, 2 H).

R$_f$ (Hex/EtOAc = 1/1): 0.54

tert-butyl(((2*R*,3*S*)-5-methoxy-3-vinyl-tetrahydrofuran-2-yl)methoxy)diphenylsilane (Method A)

Compound **158** (2.35 g, 0.61 mmol) was dissolved in 80 mL Et$_2$O. After cooling to -78° C, DIBAL-H in toluene (8.2 mL, 1.23 mmol, 1.5 M) was added and the solution stirred for 20 min. After TLC control showed complete consumption of starting material, the cooling bath was removed and the solution warmed to -50° C and then quenched with 1 mL of methanol. 10 mL of saturated KNa-tartrate solution were added and the mixture was stirred vigorously overnight. Then, the layers were separated and the aqueous layer was extracted 4 times with a total of 100 mL of Et$_2$O. The combined organic layers were washed with brine, dried over magnesium sulfate, filtered and concentrated under reduced pressure to yield lactol **205** as a colourless oil which was employed without further purification for the next step.

In a 100ml single-neck flask, the lactol was dissolved in 30 mL of MeOH and cooled to -20 °C. BF$_3$·OEt$_2$ (0.8 mL, 7.75 mmol) was added and the reaction was stirred at -20 °C for 1.5 h. The reaction was quenched by the addition of 1.2 mL of NEt$_3$. The solvent was removed under reduced pressure. Purification by column chromatography (Hex:EtOAc = 10:1 + 1% NEt$_3$) yielded **206** as a colourless oil. (2.06 g, 85%)

Mixture of diastereomers:

^{1}H NMR (250 MHz, CDCl$_3$): δ = 7.70 (m, 4 H), 7.40 (m, 6 H), 5.75 (m, 1 H), 5.01 (m, 2 H), 3.86 (m, 1 H), 3.75 (m, 1 H), 3.38 (s.1.5 H), 3.29 (s, 1.5 H), 2.94 (m, 0.5 H), 2.71 (m, 0.5 H), 2.35 (ddd, *J* = 13.4, 9.6, 5.4 Hz, 0.5 H), 2.09 (dd, *J* = 12.6, 7.1 Hz, 0.5 H), 1.86 (ddd, *J* = 12.6, 11.4, 5.0 Hz, 0.5 H), 1.74 (ddd, *J* = 13.4, 7.0, 2.8 Hz, 0.5 H), 1.07 (s, 9 H).

^{13}C NMR (63 MHz, CDCl$_3$): δ = 139.7 (CH), 138.6 (CH), 136.1 (CH), 136.08 (CH), 136.03 (CH), 135.9 (CH), 134.1 (C), 134.0 (C), 130.0 (CH), 128.0 (CH), 116.4 (CH$_2$), 115.9 (CH$_2$), 105.9 (CH), 105.0 (CH), 85.6 (CH), 83.5 (CH), 66.4 (CH$_2$), 64.8 (CH$_2$), 55.3 (CH$_3$), 54.8 (CH$_3$), 44.6 (CH), 43.8 (CH), 40.2 (CH$_2$), 39.8 (CH$_2$), 28.2 (CH$_3$), 27.2 (CH$_3$).

R$_f$ (Hex/EtOAc = 3/1): 0.60; 0.62

2-((Z)-3-Iodo-allyloxy)-tetrahydro-pyran

RO—⟋⟍—OR ⟶ O=⟨—OTHP ⟶ I—⟋⟍—OTHP
184: R = H   H   186   187
185: R = OTHP

A 50 mL single-neck flask was charged with Z-1,4-Butenediol (**184**) (10 g, 113.5 mmol) and camphor sulfonic acid (0.4 g) and cooled to 0 °C. DHP (23.86 g, 283.7 mmol) was added over a period of 20 minutes, and the reaction was then warmed to r.t. overnight.

Dilution with 20 mL of a mixture H:EtOAc = 10:1 followed by flash cromatography (eluent H:EtOAc = 10:1) yielded **185** as a viscous oil. (29.1 g, quant.)

From this batch, **185** (194 mg, 0.80 mmol) was dissolved in 5 mL of DCM and cooled to -78 °C. O$_3$ enriched air (generated using an Erwin Sander Labor Ozonisator Art. Nr. 301.19) was bubbled through the solution until a light blue colour persisted. After bubbling dry air through the solution for another 10 min, the reaction was quenched *via* addition of PPh$_3$ (485 mg, 1.85 mmol). After keeping the reaction mixture at 0 °C overnight, 1 g of silica gel was added and the solvent was carefully removed under vacuum. Purification of the silica adsorbed material by column chromatography (Hex:EtOAc = 10:1 to 2:1) yielded **186** (225 mg, 97%).

A 50 mL Schlenk-flask was charged with iodomethyl-triphenylphosphonium iodide (1241 mg, 2.34 mmol) and 15 mL of THF were added. At r.t., NaHMDS (2.34 mL, 2.34 mmol, 1M) was added swiftly. After stirring for 2 minutes, upon which the slurry dissolved and the solution turned deep-red, the reaction was cooled to -78 °C. HMPA

(419 mg, 0.41 mL, 2.34 mmol) followed by a solution of **186** (225 mg, 1.56 mmol) in 1 mL of THF was added and the reaction was kept at -78 °C for 30 min. After warming to r.t., 30 mL of pentane were added and the reaction stirred for 30 min. The slurry was then filtered off and the filtrate treated with sat. NH$_4$Cl solution. The mixture was then extracted 3 times with pentane. The combined organic layers were washed with NaHCO$_3$ and brine, dried over MgSO$_4$, filtered and concentrated under reduced pressure. Purification by column chromatography (Hex:EtOAc = 10:1) yielded **187** as a colourless oil. (347 mg, 83%)

^{1}H NMR (250 MHz, CDCl$_3$): δ = 6.48 (ddd, *J* = 7.8, 1.6 Hz, 1 H), 6.37 (td, *J* = 7.8, 1.6 Hz, 1 H), 4.65 (m, 1 H), 4.32 (dd, *J* = 5.2, 1.5 Hz, 1 H), 4.29 (ddd, *J* = 13.5, 5.2, 1.6 Hz, 1 H), 4.10 (ddd, *J* = 13.5, 5.8, 1.5 Hz, 1 H), 3.89 (m, 1 H), 3.53 (m, 1 H), 1.90-1.50 (6 H).

^{13}C NMR (63 MHz, CDCl$_3$): δ = 138.3 (CH), 98.5 (CH), 82.6 (CH), 70.0 (CH$_2$), 62.3 (CH$_2$), 30.5 (CH$_2$), 25.4 (CH$_2$), 19.4 (CH$_2$).

R$_f$ (Hex/EtOAc = 3/1): 0.54

(*E*)-3-Tributylstannanyl-prop-2-en-1-ol[68]

A 25 mL single-neck flask was charged with freshly destilled propargylic alcohol (1 g, 1.04 mL, 17.8 mmol), Bu$_3$SnH (6.75 g, 6.15 mL, 23.2 mmol) and AIBN (0.15 g, 0.90 mmol). The reaction mixture was then carefully heated to 80 °C and kept at this temperature for 2 h under vigorous stirring. After cooling to r.t., the reaction mixture was diluted with 5 mL of hexanes and purified by column chromatography (H:EtOAc = 20:1) to yield **188** as a colourless oil. (1.71 g, 60%).

¹H NMR (250 MHz, CDCl₃): δ = 6.19 (m, 2 H), 4.17 (dd, J = 6.2, 3.0 Hz, 2 H), 1.60-1.20 (m, 12 H), 0.95-0.85 (m, 15 H).

R_f (Hex/EtOAc = 3/1): 0.51

(2E,4Z)-6-(Tetrahydro-pyran-2-yloxy)-hexa-2,4-dien-1-ol

A 100 mL single-neck flask was charged with 35 mL of anhydrous DMF and PdCl₂(CH₃CN)₂ (50 mg, 0.19 mmol). Upon addition of a solution of **187** (2.54 g, 9.45 mmol) in 5 mL of DMF the solution turned brown. Then a solution of **188** (3.45 g, 9.93 mmol) in 2 mL of DMF was added dropwise.

After stirring at r.t. for 2 h, reaction control by TLC showed complete consumption of **187**. The reaction mixture was then poured onto 200 mL of diethyl ether. The organic layer was then washed twice with H₂O, dried over MgSO₄, filtered and concentrated under reduced pressure. Purification by column chromatography (Hex:EtOAc = 10:1 to 4:1) yielded **189** as a colourless oil. (1.30 g, 70%)

¹H NMR (250 MHz, CDCl₃): δ = 6.57 (dddd, J = 15.1, 11.2, 2.5, 1.4 Hz, 1 H), 6.16 (dd, J = 11.2, 10.9 Hz, 1 H), 5.88 (dd, J = 15.1, 5.6 Hz, 1 H), 5.60 (dd, J = 10.9, 6.7 Hz, 1 H), 4.62 (m, 1 H), 4.37 (ddd, J = 12.8, 6.3, 1.2 Hz, 1 H), 4.21 (m, 3 H), 3.89 (m, 1 H), 3.52 (m, 1 H), 1.90-1.50 (6 H).

¹³C NMR (63 MHz, CDCl₃): δ = 134.5 (CH), 131.1 (CH), 127.9 (CH), 126.2 (CH), 98.2 (CH), 63.6 (CH₂), 62.6 (CH₂), 31.0 (CH₂), 25.8 (CH₂), 19.8 (CH₂).

MS (EI) m/z 198 (M, 0.3), 115 (0.5), 96 (5), 85 (100).

HRMS(EI) calcd. for C₁₁H₁₈O₃: 198.1256, Found: 198.1256.

R_f (Hex/EtOAc = 3/1): 0.22

2-((2Z,4E)-6-Bromo-hexa-2,4-dienyloxy)-tetrahydro-pyran

Compound **189** (26 mg, 0.13 mmol) and NaBr (20 mg, 0.19 mmol) were dissolved in 1 mL of DMF. After dissolution of all components, 2,6-lutidine (22 mg, 24 μl, 21 mmol) was added and the solution was cooled to 0 °C. MsCl (23 mg, 20 μl, 20 mmol) was added and the mixture was stirred for 1 h at 0 °C and an additional hour at r.t.
The reaction mixture was poured onto 10 mL of H_2O and then extracted 3 times with Et_2O/pentane = 2:1. The organic layer was then washed with a sat. $CuSO_4$ solution, followed by brine, dried over $MgSO_4$, filtered and concentrated under reduced pressure. Purification by column chromatography (Hex:EtOAc = 10:1) yielded **189** as a colourless oil. (24 mg, 71%)

^{1}H NMR (250 MHz, CDCl$_3$): δ = 6.61 (dd, J = 14.6, 11.4 Hz, 1 H), 6.13 (dd, J = 11.1, 11.1 Hz, 1 H), 5.83 (dd, J = 14.6, 7.4 Hz, 1 H), 5.66 (dt, J = 11.1, 6.6 Hz, 1 H), 4.64 (t, J = 3.2 Hz, 1 H), 4.37 (dd, J = 12.8, 6.2 Hz, 1 H), 4.21 (dd, J = 12.8, 7.3 Hz, 2 H), 3.86 (m, 1 H), 3.50 (m, 1 H), 1.90-1.50 (m, 6 H).

^{13}C NMR (63 MHz, CDCl$_3$): δ = 130.0 (CH), 129.7 (CH), 129.3 (CH), 129.0 (CH), 98.0 (CH), 62.7 (CH$_2$), 62.3 (CH$_2$), 44.9 (CH$_2$), 30.6 (CH$_2$), 25.4 (CH$_2$), 19.4 (CH$_2$).

MS (EI) m/z 181 (M$^+$-Br, 0.5), 115 (5), 85 (100).

HRMS(EI) calcd. for C$_{11}$H$_{17}$O$_2$ (M-Br): 181.1229, Found: 181.1229.

R_f (Hex/EtOAc = 3/1): 0.54

2-((2Z,4E)-6-Chloro-hexa-2,4-dienyloxy)-tetrahydro-pyran

^{1}H NMR (250 MHz, CDCl$_3$): δ = 6.29 (m, 2 H), 5.84 (m, 2 H), 4.64 (t, J = 3.3 Hz, 1 H), 4.29 (dd, J = 13.4, 5.1 Hz, 1 H), 4.11 (d, J = 7.1 Hz, 2 H), 4.03 (dd, J = 13.4, 6.3 Hz, 1 H), 3.87 (m, 1 H), 3.51 (m, 1 H), 1.90-1.50 (m, 6 H).

MS (EI) m/z 216 (M$^+$, 0.6), 181 (0.5), 115 (6), 85 (100).

HRMS(EI) calcd. for C$_{11}$H$_{17}$O$_2$Cl: 216.0917, Found: 216.0920.

E-3-Tributyltin-propenal

Compound **188** (5.0g, 14.40 mmol) was dissolved in 50 mL of DCM and activated MnO$_2$ on charcoal (30g) was added. After stirring for 60 h, the mixture was filtered through a pad of celite and the solvent was removed under reduced pressure. The product **195** was separated from the starting material by column chromatography (Hex:EtOAc = 20:1). The remaining **188** was resubjected to the oxidation with 10g of MnO$_2$/C. After chromatographical separation, the batches of **195** were united. (4.42 g, 89 %)

^{1}H NMR (250 MHz, CDCl$_3$): δ = 9.41 (d, J = 7.5 Hz, 1 H), 7.79 (d, J = 19.2 Hz, 1 H), 6.62 (dd, J = 19.2, 7.5 Hz, 1 H), 1.60-1.20 (m, 12 H), 1.01 (t, J = 8.0 Hz, 6 H), 0.90 (t, J = 7.2 Hz, 1 H).

^{13}C NMR (63 MHz, CDCl$_3$): δ = 194.0 (CH), 163.5 (CH), 148.1 (CH), 29.3 (CH$_2$), 27.6 (CH$_2$), 14.0 (CH$_3$), 10.2 (CH$_2$).

R$_f$ (Hex/EtOAc = 3/1): 0.59

(2E,4Z)-5-Tributylstannanyl-penta-2,4-dienoic acid methyl ester

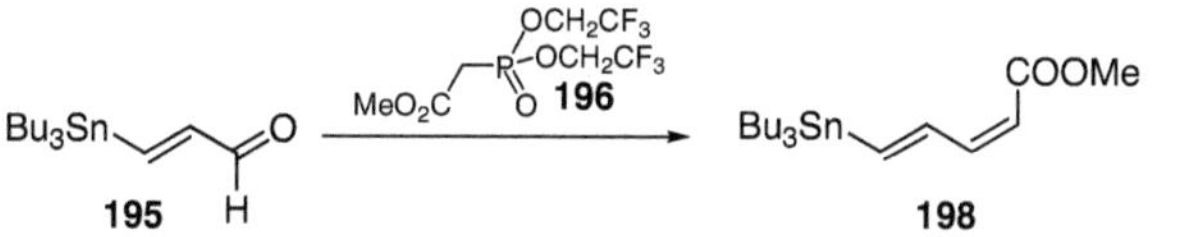

18-Crown-6 (7.93 g, 30.0 mmol) and Still-Gennari phosphonate **196** (3.82 g, 12.0 mmol) were dissolved in 120 mL of THF and cooled to -78 °C. KHMDS in toluene (20 mL, 10 mmol, 0.5M) was added dropwise, followed by **195** (3.45 g, 10.0 mmol) dissolved in 15 mL of THF. After stirring at -78 °C for 1 h, the reaction was quenched by addition of 10 mL of MeOH. After warming to r.t., the reaction mixture was concentrated under reduced pressure, followed by addition of sat. NH$_4$Cl solution. The mixture was then extracted 3 times with diethyl ether. The combined organic layers were washed with brine, dried over MgSO$_4$, filtered and concentrated under reduced pressure to yield a colourless oil which was employed without further purification for the next step. (3.94 g, 95%)

^{1}H NMR (250 MHz, CDCl$_3$): δ = 7.82 (dd, J = 19.0, 11.2 Hz, 1 H), 6.76 (d, J = 19.0, 1 H), 6.51 (dd, J = 11.4, 11.2 Hz, 1 H), 5.59 (d, J = 11.4 Hz, 1 H), 3.74 (s, 3 H), 1.60-1.20 (m, 12 H), 0.96 (t, J = 8.6 Hz, 6 H), 0.90 (t, J = 7.2 Hz, 1 H).

R$_f$ (Hex/EtOAc = 3/1): 0.74

Tributyl-[(*1E,3E*)-5-(4-methoxy-benzyloxy)-penta-1,3-dienyl]-stannane

Bu$_3$Sn—CH=CH—CH=CH—COOMe **198**   →   Bu$_3$Sn—CH=CH—CH=CH—CH$_2$OH **201**   →   Bu$_3$Sn—CH=CH—CH=CH—CH$_2$OPMB **202**

Compound **198** (4.23 g, 10.6 mmol) was dissolved in 150 mL of diethyl ether and cooled to 0 °C. DIBAl-H in toluene (17.6 mL, 26.5 mmol, 1.5M) was added dropwise, keeping the reaction temperature always below 5 °C. Stirring was continued for 30 min after the addition was completed, then the reaction was quenched by addition of 1 mL of EtOAc. 50 mL of KNa-tartrate were added and the resulting mixture was stirred vigorously at r.t. overnight.

After separation of the phases, the aqueous layer was extracted twice with diethyl ether. The combined organic layers were washed with brine, dried over MgSO$_4$, filtered and concentrated under reduced pressure. Purification by column chromatography (Hex:EtOAc = 10:1) yielded **201** as a colourless oil. (3.95 g, quant.)

^{1}H NMR (250 MHz, CDCl$_3$): δ = 6.81 (ddd, *J* = 18.5, 10.5 Hz, 1 H), 6.34 (d, *J* = 18.5 Hz, 1 H), 6.07 (dd, *J* = 11.1, 11.0 Hz, 1 H), 5.54 (dt, *J* = 11.0, 7.0 Hz, 1 H), 4.37 (m, 1 H), 1.60-1.20 (m, 12 H), 0.95-0.85 (m, 15 H).

R$_f$ (Hex/EtOAc = 3/1): 0.43

Compound **201** (3.95 g, 10.6 mmol) was dissolved in 40 mL of THF, cooled to -10 °C and NaH (0.85 g, 21.2 mmol, 60 %) was added. After stirring for 30 min, PMBCl (1.83 g, 11.7 mmol) and TBAI (0.1 g) were added. Then the solution was warmed to 40 °C and stirred at this temperature overnight.

After cooling to r.t, the reaction was quenched by addition of sat. NH$_4$Cl solution. The mixture was then extracted 4 times with hexanes. The combined organic layers were washed with NaHCO$_3$ and brine, dried over MgSO$_4$, filtered and concentrated under reduced pressure. Purification by column chromatography (Hex:EtOAc = 20:1) yielded **202** as a colourless oil. (4.18 g, 80 %)

^{1}H NMR (250 MHz, CDCl$_3$): δ = 7.28 (d, 8.7 Hz, 1 H), 6.88 (d, *J* = 8.7 Hz, 1 H), 6.77 (ddd, *J* = 18.6, 11.0, 1.4 Hz, 1 H), 6.31 (d, *J* = 18.6 Hz, 1 H), 5.52 (dt, *J* = 11.0, 6.7 Hz,

1 H), 4.46 (s, 2 H), 4.21 (dd, J = 6.7, 1.4 Hz, 2 H), 3.81 (s, 3 H), 1.60-1.20 (m, 12 H), 0.91 (t, J = 8.0 Hz, 6 H), 0.89 (t, J = 7.2 Hz, 9 H).

^{13}C NMR (63 MHz, CDCl$_3$): δ = 141.8 (CH), 137.5 (CH), 135.0 (CH), 129.8 (CH), 126.4 (CH), 114.2 (CH), 72.1 (CH$_2$), 66.0 (CH$_2$), 55.6 (CH$_3$), 29.5 (CH$_2$), 27.6 (CH$_2$), 14.1 (CH$_3$), 10.0 (CH$_2$).

R$_f$ (Hex/EtOAc = 3/1): 0.53

tert-Butyl-[(2*R*,3*S*)-3-((*Z*)-2-iodo-vinyl)-5-methoxy-tetrahydro-furan-2-ylmethoxy]-diphenyl-silane

Compound **206** (200 mg, 0.50 mmol) was dissolved in 20 mL of CH$_2$Cl$_2$ and cooled to -78 °C. O$_3$ enriched air (generated using an Erwin Sander Labor Ozonisator Art. Nr. 301.19) was bubbled through the solution until a light blue colour persisted. After bubbling dry air through the solution for another 10 min, the reaction was quenched *via* addition of PPh$_3$ (304 mg, 1.16 mmol). The reaction mixture was kept at 0 °C overnight, 1 g of silica gel was added and the solvent was carefully removed under vacuum. Flash chromatography using Hex:EtOAc = 7:1 yielded **207** sufficiently pure for the next step. A 50 mL Schlenk-flask was charged with iodomethyl-triphenylphosphonium iodide (321 mg, 0.61 mmol) and 20 mL of THF were added. At r.t., NaHMDS (0.61 mL, 0.61 mmol, 1M) was added swiftly. After stirring for 1 min, upon which the slurry dissolved and the solution turned deep-red, the reaction was cooled to -78 °C. HMPA (108 mg, 0.11 mL, 0.61 mmol), followed by a solution of **186** (225 mg, 1.56 mmol), was added and the reaction was kept at -78 °C for 30 min. After warming to r.t., the reaction was quenched by addition of 2 mL of H$_2$O – then the mixture was concentrated under reduced pressure. The residue was then treated with sat. NH$_4$Cl solution and then extracted 3 times with EtOAc. The combined organic layers were washed with sat.

NaHCO$_3$ and brine, dried over MgSO$_4$, filtered and concentrated under reduced pressure. Purification by column chromatography (Hex:EtOAc = 4:1) yielded **187** as a colourless oil. (179 mg, 83%, mixture of diastereomers)

208a:

^{1}H NMR (250 MHz, CDCl$_3$): δ = 7.72 (m, 4 H), 7.39 (m, 6 H), 6.29 (d, J = 7.3 Hz, 1 H), 6.10 (dd, J = 8.7, 7.3 Hz, 1 H), 5.01 (d, J = 4.8 Hz, 1 H), 3.97 (ddd, J = 8.7, 5.3, 4.5 Hz, 1 H), 3.79 (dd, J = 10.3, 4.5 Hz, 1 H), 3.73 (dd, J = 10.3, 5.3 Hz, 1 H), 3.29 (s, 3 H), 3.27 (m, 1 H), 2.23 (dd, J = 12.8, 7.3 Hz, 1 H), 1.78 (ddd, J = 12.8, 11.0, 5.0 Hz, 1 H), 1.07 (s, 9 H).

^{13}C NMR (63 MHz, CDCl$_3$): δ = 141.6 (CH), 136.13 (CH), 136.08 (CH), 135.2 (CH), 134.0 (C), 130.0 (CH), 128.05 (CH), 128.01 (CH), 105.3 (CH), 85.4 (CH), 83.8 (CH), 66.8 (CH$_2$), 55.0 (CH$_3$), 45.5 (CH), 39.2 (CH$_2$), 27.3 (CH$_3$), 19.6 (C).

R$_f$ (Hex/EtOAc = 3/1): 0.57

208b:

^{1}H NMR (250 MHz, CDCl$_3$): δ = 7.73 (m, 4 H), 7.42 (m, 6 H), 6.30 (dd, J = 8.7, 7.3 Hz, 1 H), 6.21 (d, J = 7.3 Hz, 1 H), 5.09 (dd, J = 6.5, 4.3, 3.4 Hz, 1 H), 3.81 (dd, J = 11.0, 3.4 Hz, 1 H), 3.75 (dd, J = 11.0, 4.3 Hz, 1 H), 3.37 (s, 3 H), 3.15 (m, 1 H), 2.44 (ddd, J = 13.4, 10.0, 5.2 Hz, 1 H), 1.71 (ddd, J = 13.4, 5.3, 2.0 Hz, 1 H), 1.08 (s, 9 H).

R$_f$ (Hex/EtOAc = 3/1): 0.67

((2R,3S,5S)-3-Ethynyl-5-methoxy-tetrahydro-furan-2-yl)-methanol

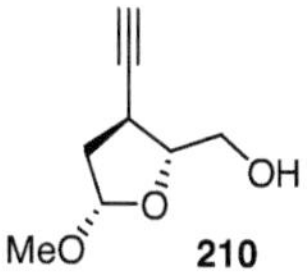

MeO — **210** — OH

¹H NMR (250 MHz, CDCl₃): δ = 5.01 (d, *J* = 4.8 Hz, 1 H), 4.19 (ddd, *J* = 8.2, 3.5, 3.2 Hz, 1 H), 3.82 (bd, *J* = 12.1 Hz, 1 H), 3.60 (bd, *J* = 12.1 Hz, 1 H), 3.36 (s, 3 H), 3.17 (m, 1 H), 2.31 (dd, *J* = 12.8, 7.5 Hz, 1 H), 2.11 (m, 1 H), 2.10 (d, 2.3 Hz, 1 H), 1.07 (s, 9 H).

¹³C NMR (63 MHz, CDCl₃): δ = 105.7 (CH), 105.4 (CH), 86.5 (CH), 85.6 (CH), 63.5 (CH₂), 55.5 (CH₃), 40.8 (CH₂), 28.8 (CH).

IR [cm⁻¹]: 3424, 3300, 3071, 3050, 2996, 2931, 2122, 1472, 1428, 1390, 1363, 1331, 1302, 1267, 1204, 1158, 1111, 1062, 1017.

R_f (Hex/EtOAc = 3/1): 0.13

((2R,3S)-5-Methoxy-3-vinyl-tetrahydro-furan-2-yl)-methanol

MeO — **206** — OTBDPS → MeO — **211** — OH

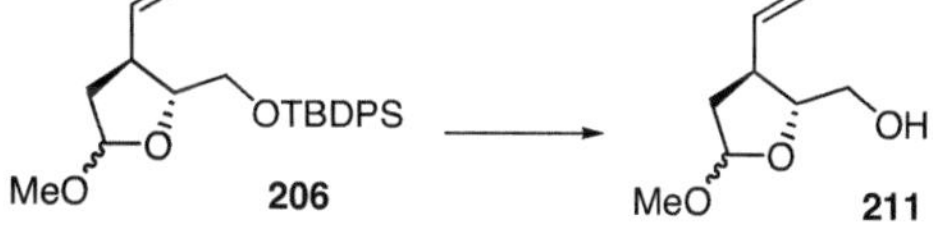

Compound **206** (783 mg, 1.97 mmol) was dissolved in 2 mL of THF and a 1M solution of TBAF in THF (4.93 mL, 4.93 mmol, 1.0 M) was added. After stirring at r.t. for 3 h, the solvent was removed under reduced pressure, and the residue was purified by column chromatography (Hex:EtOAc = 10:1 to 2:1) to give **211** as a colourless oil. (300 mg, 96 %).

¹H NMR (250 MHz, CDCl₃): δ = 5.74 (m, 1 H), 5.15-4.96 (m, 3 H), 3.92 (ddd, J = 8.2, 4.4, 2.7 Hz, 0.5 H), 3.86-3.73 (m, 1.5 H) 3.55 (m, 0.5 H), 3.37 (2 × s, 3 H), 2.98 (m, 0.5 H), 2.66 (m, 0.5 H), 2.38 (ddd, J = 13.5, 9.8, 5.4 Hz, 0.5 H), 2.13 (dd, J = 12.8, 7.3 Hz, 0.5 H), 1.87 (ddd, J = 12.8, 11.1, 5.0 Hz, 0.5 H), 1.76 (ddd, J = 13.5, 7.7, 2.9 Hz, 0.5 H).

¹³C NMR (63 MHz, CDCl₃): δ = 138.9 (CH), 138.4 (CH), 117.0 (CH₂), 116.8 (CH₂), 105.7 (CH), 105.5 (CH), 85.8 (CH), 82.7 (CH), 63.8 (CH₂), 62.7 (CH₂), 55.5 (CH₃), 55.3 (CH₃), 44.2 (CH), 42.3 (CH), 40.4 (CH₂), 40.1 (CH₂).

R_f (Hex/EtOAc = 3/1): 0.11

[(2R,3S)-3-((Z)-2-Iodo-vinyl)-5-methoxy-tetrahydro-furan-2-yl]-methanol

Compound **208** (84 mg, 0.16 mmol) was dissolved in 1.2 mL of THF and a 7% solution of HF·pyridine (0.8 mL, 2.80 mmol, 7 %) was added. After stirring at r.t. for 60 h, the solution was poured onto sat. NaHCO₃ solution and subsequently extracted 4 times with DCM. The solvent was removed under reduced pressure, and the residue was treated twice with toluene, which was also removed under reduced pressure. Purification by column chromatography (Hex:EtOAc = 10:1 to 2:1 + 5 % NEt₃) yielded **216** as a colourless oil. (37 mg, 82 %).

216a:
¹H NMR (250 MHz, CDCl₃): δ = 6.27 (m, 2 H), 5.09 (dd, J = 5.3, 2.1 Hz, 1 H), 3.91 (ddd, J = 7.5, 4.8, 2.8 Hz, 1 H), 3.82 (dd, J = 12.0, 2.8 Hz, 1 H), 3.62 (dd, J = 12.0, 4.8 Hz, 1 H), 3.36 (s, 3 H), 3.02 (s, 3 H), 2.42 (ddd, J = 13.6, 10.0, 5.3 Hz, 1 H), 1.74 (ddd, J = 13.6, 6.0, 2.2 Hz, 1 H).

¹³C NMR (63 MHz, CDCl₃): δ = 142.5 (CH), 105.8 (CH), 83.6 (CH), 82.8 (CH), 63.6 (CH₂), 63.6 (CH₂), 55.4 (CH₃), 45.0 (CH), 39.2 (CH₂).

R_f (Hex/EtOAc = 3/1): 0.12

216b:

¹H NMR (250 MHz, CDCl₃): δ = 6.34 (d, *J* = 7.5 Hz, 1 H), 6.11 (dd, *J* = 9.0, 7.5 Hz, 1 H), 5.02 (d, *J* = 5.0 Hz, 1 H), 3.98 (ddd, *J* = 7.6, 4.8, 2.8 Hz, 1 H), 3.76 (bd, *J* = 11.6 Hz, 1 H), 3.59 (d, *J* = 11.6 Hz, 1 H), 3.39 (s, 3 H), 3.34 (m, 1 H), 2.25 (dd, *J* = 13.0, 7.5 Hz, 1 H), 1.86 (ddd, *J* = 13.0, 10.5, 5.0 Hz, 1 H).

¹³C NMR (63 MHz, CDCl₃): δ = 141.6 (CH), 105.7 (CH), 85.6 (CH), 84.4 (CH), 64.6 (CH₂), 55.5 (CH₃), 43.9 (CH), 39.6 (CH₂).

R_f (Hex/EtOAc = 3/1): 0.22

(Z)-3-[(2R,3S,5S)-3-((Z)-2-Iodo-vinyl)-5-methoxy-tetrahydro-furan-2-yl]-acrylic acid ethyl ester

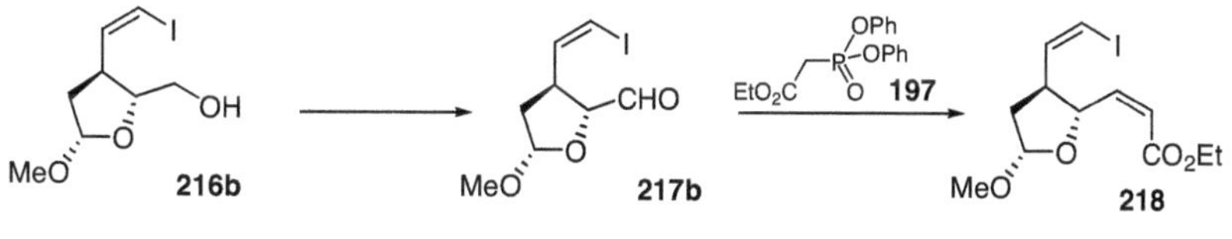

Compound **216b** (32 mg, 0.11 mmol) was dissolved in 1 mL of DCM. 1 mL of DMSO was added and the mixture was cooled to 0 °C, when NEt₃ (34 mg, 47 µl, 0.34 mmol) followed by a solution of SO₃·pyridine (54 mg, 0.34 mmol) in 1 mL of DMSO. After stirring at r.t. for 3 h, the reaction was quenched by addition of H₂O. The mixture was then extracted 4 times with hexanes:Et₂O = 1:1. The combined organic layers were washed with brine, dried over MgSO$_4$, filtered and concentrated under reduced pressure to yield the crude aldehyde **217b** which was employed without further purification for the next step.

Ando phosphonate **197** (61 mg, 0.19 mmol) was dissolved in 1.5 mL of THF and cooled to 0 °C. NaH-suspension (8 mg, 0.19 mmol, 60%) was added and the mixture stirred at 0 °C for 45 min. After cooling to -78 °C, a solution of aldehyde **217b** in 1 mL of THF was added and stirring continued at this temperature for 1.5 h. After warming to r.t., the reaction was quenched by addition of H$_2$O. The mixture was then extracted 3 times with diethyl ether. The combined organic layers were washed with brine, dried over MgSO$_4$, filtered and concentrated under reduced pressure. Purification by column chromatography (Hex:EtOAc = 20:1 + 1% NEt$_3$) yielded **218** as a colourless oil. (17 mg, 76%)

^{1}H NMR (400 MHz, CDCl$_3$): δ = 6.39-6.20 (m, 3 H), 5.85 (d, J = 11.9 Hz, 1 H), 5.58 (dd, J = 9.8, 8.9 Hz, 1 H), 5.06 (d, J = 4.8 Hz, 1 H), 3.70 (s, 3 H), 3.40 (s, 3 H), 3.26 (m 1 H), 2.29 (dd, J = 12.6, 7.1 Hz, 1 H), 1.88 (dd, J= 12.6, 11.0, 5.0 Hz, 1 H).

^{13}C NMR (100 MHz, CDCl$_3$): δ = 185.5 (C), 149.6 (CH), 140.8 (CH), 120.8 (CH), 105.8 (CH), 84.1 (CH), 78.9 (CH), 55.1 (CH$_3$), 51.7 (CH$_3$), 49.6 (CH), 39.6 (CH$_2$).

R$_f$ (Hex/EtOAc = 1/1): 0.68

(Z)-3-{(2S,3S,5S)-5-Methoxy-3-[(1Z,3E,5Z)-7-(4-methoxy-benzyloxy)-hepta-1,3,5-trienyl]-tetrahydro-furan-2-yl}-acrylic acid ethyl ester

A 10 mL Schlenk flask was charged with 1 mL of DMSO, which was degassed in 3 freeze/pump/thaw cycles. Pd$_2$dba$_3$ (1 mg, 1.04 μmol), P(o-furyl)$_3$ (1 mg, 4.14 μmol) and

CuI (5 mg, 0.03 mmol) were added, followed by a solution of **218** (7 mg, 0.03 mmol) in 0.5 mL of DMSO. Then, a solution of **202** (15 mg, 0.10 mmol) in 0.5 mL of DMSO was added and the solution stirred at 40 °C overnight.

The reaction mixture was diluted with EtOAc and and then washed twice with a 1:1 mixture of brine and 5 % NH_4OH. The aqueous layer was extracted back once with EtOAc, then the combined organic layers were washed with brine, dried over $MgSO_4$, filtered and concentrated under reduced pressure. Purification by column chromatography (Hex:EtOAc = 4:1 to 2:1) yielded **219** as a colourless oil. (5 mg, 60%)

^{1}H NMR (600 MHz, $CDCl_3$): δ = 7.24 (d, J = 8.6 Hz, 2 H), 6.86 (d, J = 8.6 Hz, 2 H), 6.35 (m, 2 H), 6.17 (dd, J = 10.1, 10.1 Hz, 1 H), 6.10 (dd, J = 11.7, 8.6 Hz, 1 H), 6.03 (dd, J = 10.5, 10.5 Hz, 1 H), 5.82 (d, J = 11.6 Hz, 1 H), 5.59 (dt, J = 11.2, 6.7 Hz, 1 H), 5.55 (dd, J = 10.4, 10.4 Hz, 1 H), 5.41 (dd, J = 8.4, 8.4 Hz, 1 H), 5.09 (dd, J = 5.5, 3.2 Hz, 1 H), 4.42 (s, 2 H), 4.13 (d, J = 6.7 Hz, 1 H), 4.05 (q, J = 7.1 Hz, 2 H), 3.36 (s, 3 H), 2.97 (m, 1 H), 2.45 (ddd, J = 13.8, 9.0, 5.2 Hz, 1 H), 1.76 (ddd, J = 13.8, 8.3, 3.1 Hz, 1 H), 1.18 (J = 7.1 Hz, 3 H).

R_f (Hex/EtOAc = 3/1): 0.21

(Z)-3-[(2S,3S,5S)-3-((1Z,3E,5Z)-7-Hydroxy-hepta-1,3,5-trienyl)-5-methoxy-tetrahydro-furan-2-yl]-acrylic acid ethyl ester

A 10 mL Schlenk flask was charged with 1.5 mL of DMSO, which was degassed in 3 freeze/pump/thaw cycles. Pd_2dba_3 (2.2 mg, 2.41 μmol), P(o-furyl)$_3$ (2.2 mg, 9.65 μmol) and CuI (12 mg, 0.06 mmol) were added, followed by a solution of **218** (17 mg, 0.05 mmol) in 0.5 mL of DMSO. Then, a solution of **201** (36 mg, 0.10 mmol) in 0.5 mL of DMSO was added and the solution stirred at 40 °C overnight.

The reaction mixture was diluted with EtOAc and and then washed twice with a 1:1 mixture of brine and 5 % NH₄OH. The aqueous layer was extracted back once with EtOAc, then the combined organic layers were washed with brine, dried over MgSO₄, filtered and concentrated under reduced pressure. Purification by column chromatography (Hex:EtOAc = 4:1 to 2:1) yielded **220** as a colourless oil. (8 mg, 54%)

^{1}H NMR (400 MHz, CDCl₃): δ = 6.46 (m, 2 H), 6.20-6.09 (m, 3 H), 5.80 (d, J = 11.6 Hz, 1 H), 5.64 (dt, J = 10.9, 7.0 Hz, 1 H), 5.47 (m, 2 H), 5.04 (d, J = 4.8 Hz, 1 H), 4.32 (bd, J = 7.0 Hz, 2 H), 4.05 (q, J = 7.1 Hz, 2 H), 3.40 (s, 3 H), 3.33 (m, 1 H), 2.16 (dd, J = 12.3, 6.8 Hz, 1 H), 1.89 (ddd, J = 12.8, 11.6, 4.9 Hz, 1 H), 1.18 (t, J = 7.1 Hz, 3 H).

^{13}C NMR (100 MHz, CDCl₃): δ = 148.9 (CH), 132.1 (CH), 131.2 (CH), 130.9 (CH), 130.4 (CH), 129.9 (CH), 128.4 (CH), 121.4 (CH), 105.8 (CH), 80.1 (CH), 60.6 (CH₂), 59.2 (CH₂), 55.0 (CH₃), 43.2 (CH), 41.3 (CH₂), 14.6 (CH₃).

R$_f$ (Hex/EtOAc = 1/1): 0.23

(E)-3-[(2S,3S,5S)-3-((*1Z,3E,5E*)-7-Hydroxy-hepta-1,3,5-trienyl)-5-methoxy-tetrahydro-furan-2-yl]-acrylic acid ethyl ester

221

^{1}H NMR (400 MHz, CDCl₃): δ = 6.42 (dd, J = 14.5, 11.2 Hz, 1 H), 6.27 (dd, J = 15.0, 10.6 Hz, 1 H), 6.17 (dd, J = 14.5, 10.7 Hz, 1 H), 6.15 (dd, J = 11.8, 8.7 Hz, 1 H), 6.06 (dd, J = 11.0, 11.0 Hz, 1 H), 5.83 (dt, J = 15.2, 6.1 Hz, 1 H), 5.78 (dd, J = 11.9, 1.0 Hz, 1 H), 5.46 (dd, J = 9.0, 9.0 Hz, 1 H), 5.41 (dd, J = 10.7, 10.3 Hz, 1 H), 5.01 (d, J = 4.6 Hz, 1 H), 4.18 (bs, OH), 4.10 (q, J = 7.2 Hz, 2 H), 3.37 (s, 3 H), 3.29 (m, 1 H), 2.14 (dd, J = 12.8, 6.4 Hz, 1 H), 1.86 (ddd, J = 12.8, 11.5, 4.7 Hz, 1 H), 1.23 (t, J = 7.2 Hz, 3 H).

^{13}C NMR (100 MHz, CDCl$_3$): δ = 148.9 (CH), 133.1 (CH), 132.8 (CH), 131.9 (CH), 131.4 (CH), 130.9 (CH), 128.5 (CH), 121.4 (CH), 105.8 (CH), 80.0 (CH), 63.8 (CH$_2$), 60.6 (CH$_2$), 55.1 (CH$_3$), 43.1 (CH), 41.5 (CH$_2$), 14.6 (CH$_3$).

R$_f$ (Hex/EtOAc = 1/1): 0.25

(E)-3-[(2S,3S,5R)-3-((Z)-2-Iodo-vinyl)-5-methoxy-tetrahydro-furan-2-yl]-acrylic acid ethyl ester

Benzoic acid (16 mg, 0.13 mmol), Dess-Martin periodinane **112** (34 mg, 0.08 mmol) and stabilised Wittig reagent **138** were dissolved in 0.5 mL of DCM and 0.09 mL of DMSO. Compound **216a**, dissolved in 0.5 mL of DCM was added and the mixture stirred at r.t. overnight.

The reaction was quenched by addition of a 1:1 mixture of 1M Na$_2$S$_2$O$_3$ and sat. NaHCO$_3$. After stirring for 1 h, the mixture was extracted 4 times with diethyl ether. The combined organic layers were washed with sat. NaHCO$_3$ and brine, dried over MgSO$_4$, filtered and concentrated under reduced pressure. Purification by column chromatography (Hex:EtOAc = 10:1 + 1% NEt$_3$) yielded **222** as a colourless oil. (18 mg, 76%)

^{1}H NMR (250 MHz, CDCl$_3$): δ = 6.94 (dd, 15.7, 5.1 Hz, 1 H), 6.36 (d, J = 7.5 Hz, 1 H), 6.29 (dd, J = 8.1, 7.7 Hz, 1 H), 6.05 (dd, J = 15.5, 1.6 Hz, 1 H), 5.13 (dd, J = 5.5, 2.3 Hz, 1 H), 4.40 (ddd, J = 7.3, 5.4, 1.7 Hz, 1 H), 4.20 (q, J = 7.2 Hz, 2 H), 3.37 (s, 3 H), 2.94 (m, 1 H), 2.46 (ddd, J = 13.5, 9.5, 5.4 Hz, 1 H), 1.75 (ddd, J = 13.5, 6.4, 2.3 Hz, 1 H), 1.28 (t, J = 7.2 Hz, 3 H).

^{13}C NMR (63 MHz, CDCl$_3$): δ = 166.5 (C), 145.4 (CH), 144.1 (CH), 121.8 (CH), 105.8 (CH), 84.7 (CH), 81.0 (CH), 60.8 (CH$_2$), 55.6 (CH$_3$), 49.7 (CH), 38.7 (CH$_2$), 14.6 (CH$_3$).

R$_f$ (Hex/EtOAc = 1/1): 0.67

(E)-3-[(2S,3S,5R)-3-((1Z,3E,5Z)-7-Hydroxy-hepta-1,3,5-trienyl)-5-methoxy-tetrahydro-furan-2-yl]-acrylic acid ethyl ester

A 10 mL Schlenk flask was charged with 1.5 mL of DMSO, which was degassed in 3 freeze/pump/thaw cycles. Pd$_2$dba$_3$ (2.3 mg, 2.55 µmol), P(o-furyl)$_3$ (2.3 mg, 10.02 µmol) and CuI (13 mg, 0.07 mmol) were added, followed by a solution of **222** (18 mg, 0.05 mmol) in 0.5 mL of DMSO. Then, a solution of **201** (38 mg, 0.10 mmol) in 0.5 mL of DMSO was added and the solution stirred at 40 °C overnight.

The reaction mixture was diluted with EtOAc and and then washed twice with a 1:1 mixture of brine and 5 % NH$_4$OH. The aqueous layer was extracted back 4 times with EtOAc, then the combined organic layers were washed with brine, dried over MgSO$_4$, filtered and concentrated under reduced pressure. Purification by column chromatography (Hex:EtOAc = 4:1 to 2:1) yielded **223** as a colourless oil. (11.5 mg, 73%)

^{1}H NMR (250 MHz, CDCl$_3$): δ = 6.87 (dd, J = 15.5, 5.3 Hz, 1 H), 6.50 (dd, J = 14.5, 10.9 Hz, 1 H), 6.37 (dd, J = 14.3, 11.1 Hz, 1 H), 6.37 (dd, J = 10.3, 10.3 Hz, 1 H), 6.14 (dd, J = 10.3, 10.3 Hz, 1 H), 6.12 (dd, J = 10.8, 10.8 Hz, 1 H), 6.04 (dd, J = 15.8, 1.4 Hz, 1 H), 5.65 (dt, J = 11.1, 7.0 Hz, 1 H), 5.46 (dd, J = 10.2, 10.2 Hz, 1 H), 5.12 (dd, J = 5.5, 3.0 Hz, 1 H), 4.33 (dd, J = 6.9, 1.4 Hz, 2 H), 4.18 (q, J = 7.1 Hz, 2 H), 3.38 (s, 3

H), 3.00 (m, 1 H), 2.47 (ddd, J = 13.6, 9.5, 5.5 Hz, 1 H), 1.74 (ddd, J = 13.6, 7.8, 2.9 Hz, 1 H), 1.27 (t, J = 7.1 Hz, 3 H).

^{13}C NMR (63 MHz, CDCl$_3$): δ = 145.4 (CH), 131.7 (CH), 131.1 (CH), 131.0 (CH), 130.8 (CH), 129.5 (CH), 129.1 (CH), 121.9 (CH), 105.8 (CH), 81.5 (CH), 60.8 (CH$_2$), 59.3 (CH$_2$), 55.7 (CH$_3$), 44.0 (CH), 40.7 (CH$_2$), 14.6 (CH$_3$).

R$_f$ (Hex/EtOAc = 1/1): 0.31

11.3 Third Generation Approach

tert-butyl(((2*R*,3*S*,5*R*)-5-methoxy-3-vinyl-tetrahydrofuran-2-yl)methoxy)diphenylsilane (Method B)

Compound **158** (6.05 g, 14.82 mmol) was dissolved in 200 mL Et$_2$O and transferred into a 500 mL one-neck flask. After cooling to -78° C, DIBAL-H in toluene (17.0 mL, 25.44 mmol, 1.5 M) was added and the solution stirred for one hour. After TLC control showed complete consumption of starting material, the cooling bath was removed and the solution warmed to -50° C and then quenched with 2 mL of methanol. 70 mL of saturated KNa-tartrate solution was added and the mixture was stirred vigorously overnight. Then, the layers were separated and the aqueous layer was extracted 4 times with a total of 200 mL of Et$_2$O. The combined organic layers were washed with brine, dried over magnesium sulfate, filtered and concentrated under reduced pressure to yield a colourless oil (5.47g), which was employed without further purification for the next step.

In a 100ml single-neck flask, the lactol was dissolved in 30 mL of CH$_2$Cl$_2$; then, Ag$_2$O (6.63 g, 28.60 mmol) and MeI (10.15 g, 4.45 mL, 71.49 mmol) were added. The resulting suspension was heated to reflux for 48 h. After cooling to r.t., the reaction mixture was filtered through a pad of celite and concentrated under reduced pressure. Purification by column chromatography (Hex:EtOAc = 10:1 + 1% NEt$_3$) yielded **206a** as a colourless oil (5.00 g, 85 %).

^{1}H NMR (400 MHz, CDCl$_3$): δ = 7.72 (m, 4 H), 7.42 (m, 6 H), 5.80 (ddd, *J* = 17.0, 10.1, 8.6 Hz, 1 H), 5.08 (dd, *J* = 5.4, 2.7 Hz, 1 H), 5.0 (dd, *J* = 17.0, 1.5 Hz, 1 H), 4.97 (dd, *J* = 10.1, 1.5 Hz, 1 H), 3.88 (ddd, *J* = 7.7, 4.6, 3.4 Hz, 1 H), 3.83 (dd, *J* = 11.1, 3.4 Hz, 1 H), 3.73 (dd, *J* = 11.1, 4.6 Hz, 1 H), 3.41 (s, 3 H), 2.73 (dddd, *J* = 10.1, 9.6, 7.7,

7.0 Hz, 1 H), 2.37 (ddd, J = 13.4, 9.6, 5.4 Hz, 1 H), 1.76 (ddd, J = 13.4, 7.0, 2.7 Hz, 1 H).

^{13}C NMR (100 MHz, CDCl$_3$): δ =139.7 (CH), 136.1 (CH), 136.0 (CH), 134.1 (C), 130.0 (CH), 129.9 (CH), 128.0 (CH), 127.9 (CH), 115.9 (CH$_2$), 105.6 (CH), 83.4 (CH), 64.8 (CH$_2$), 55.3 (CH$_3$), 44.6 (CH), 39.8 (CH$_2$), 27.2 (CH$_3$), 19.7 (C).

HRMS(EI) calcd. for C$_{24}$H$_{32}$O$_3$Si: 396.2121, Found: 396.2128.

R$_f$ (Hex/EtOAc = 3/1): 0.61

tert-Butyl(((2*R*,3*S*,5*R*)-5-methoxy-3-((*1Z*,3*Z*)-4-(tributylstannyl)buta-1,3-dienyl)-tetrahydrofuran-2-yl)methoxy)-diphenylsilane

Compound **206a** (4.82 g, 12.16 mmol) was dissolved in 150 mL of CH$_2$Cl$_2$ and cooled to -78 °C. O$_3$ enriched air (generated using an Erwin Sander Labor Ozonisator Art. Nr. 301.19) was bubbled through the solution until a light blue colour persisted. After bubbling dry air through the solution for another 10 min, the reaction was quenched via addition of PPh$_3$ (7.34g, 28.0 mmol). After keeping the reaction mixture at 0 °C overnight, 15 g of silica gel were added and the solvent was carefully removed under vacuum. Purification of the silica adsorbed material by column chromatography using Hex:EtOAc = 10:1 to 7:1 as an eluent yielded **207a** as a colourless oil. (4.00 g, 83 %)

^{1}H NMR (250 MHz, CDCl$_3$): δ = 9.70 (d, J = 2.1 Hz, 1 H), 6.67 (m, 4 H), 7.42 (m, 6 H), 5.08 (dd, J = 3.2, 2.3 Hz, 1 H), 4.44 (dd, J = 8.9, 4.3 Hz, 1 H), 3.77 (d, J = 4.3 Hz, 2 H) 3.29 (s, 3 H), 2.94 (m 1 H), 2.27 (m, 2 H), 1.06 (s, 9 H).

^{13}C NMR (75 MHz, CDCl$_3$): δ = 201.7 (CH), 135.6 (CH), 135.5 (CH), 133.6 (C), 129.8 (CH), 129.7 (CH), 127.8 (CH), 127.7 (CH), 104.6 (CH), 78.3 (CH), 65.3 (CH$_2$), 54.2 (CH$_3$), 51.8 (CH), 34.2 (CH$_2$), 26.8 (CH$_3$), 19.2 (C).

R$_f$ (Hex/EtOAc = 10/1): 0.47

[(2R,3S,5R)-3-((Z)-2-Iodo-vinyl)-5-methoxy-tetrahydro-furan-2-yl]-methanol (Method B)

Compound **208a** (555 mg, 1.06 mmol) was dissolved in 5 mL of MeOH and NH$_4$F (393 mg, 10.62 mmol) was added.

After stirring at r.t. overnight, the solvent was removed under reduced pressure, the residue taken up in diethyl ether and sat. NaHCO$_3$. The mixture was then extracted 4 times with diethyl ether. The combined organic layers were washed with brine, dried over MgSO$_4$, filtered and concentrated under reduced pressure. Purification by column chromatography (Hex:EtOAc = 10:1 to 4:1 + 5% NEt$_3$) yielded **216a** as a colourless oil. (247 mg, 82%)

Spectroscopic analysis was identical to that of **216a** obtained by Method A

(Diethoxy-phosphoryl)-acetic acid (2E,4Z)-5-tributylstannanyl-penta-2,4-dienyl ester

Phosponoacetic acid (**233**) (420 mg, 2.14 mmol) was dissolved in 3 mL of CH_2Cl_2 and cooled to 0 °C. Oxalylchloride (408 mg, 0.28 mL, 3.22 mmol) was added slowly. After stirring at 0° C for 20 min, the reaction mixture was warmed to room temperature and stirring continued for 2 h. Then the volatile components were removed under reduced pressure and the acid chloride was used without any further purification.

Compound **201** (400 mg, 1.07 mmol) was dissolved in 7 mL CH_2Cl_2 and cooled to 0 °C. Pyridine (254 mg, 0.26 mL, 3.22 mmol) and DMAP (20 mg) were added, then the acid chloride (dissolved in 3 mL CH_2Cl_2) was added slowly *via* cannula. The reaction was warmed to room temperature and stirred overnight.

The reaction was quenched by addition of sat. $NaHCO_3$ solution. After separation of the organic layer, the aqueous layer was subsequently extracted 4 times with a total of 200 mL of CH_2Cl_2. The combined organic layers were washed with brine, dried over magnesium sulfate, filtered and concentrated under reduced pressure. Purification by column chromatography (Hex:EtOAc = 2:1 + 1% NEt_3) yielded **235** as a colourless oil. (442 mg, 75%)

^1H NMR (250 MHz, $CDCl_3$): δ = 6.77 (ddd, J = 18.5, 10.5, 1.1 Hz, 1 H), 6.38 (d, J = 18.5 Hz, 1 H), 6.13 (dd, J = 11.4, 10.7 Hz, 1 H), 5.44 (dt, J = 10.9, 7.1 Hz, 1 H), 4.84 (dd, J = 7.1, 1.1 Hz, 2 H), 4.16 (dq, J = 7.8, 7.2 Hz, 4 H), 2.97 (d, J = 21.5 Hz, 2 H), 1.60-1.20 (m, 18 H), 0.95-0.85 (m, 15 H).

^{13}C NMR (63 MHz, $CDCl_3$): δ = 165.6 (C), 140.4 (CH), 139.2 (CH), 135.8 (CH), 121.9 (CH), 66.7 (d, J =6.0 Hz, CH_2), 61.5 (CH_2), 34.3 (d, J = 113 Hz, CH_2), 29.0 (CH_2), 27.1 (CH_2), 16.3 (d, J = 6 Hz, CH_3), 13.6 (CH_3), 9.5 (CH_2).

R_f (Hex/EtOAc = 3/1): 0.09

(Z)-3-[(2S,3S,5R)-3-((Z)-2-Iodo-vinyl)-5-methoxy-tetrahydro-furan-2-yl]-acrylic acid (2Z,4Z)-5-tributylstannanyl-penta-2,4-dienyl ester

Compound **216** (92 mg, 0.32 mmol) was dissolved in 1 mL of CH$_3$CN and IBX (272 mg, 0.97 mmol) was added. The suspension was heated to reflux for 30 min; after cooling to r.t., the slurry was filtered and diluted with 1 mL of CH$_3$CN.

In a separate flask, compound **235** (294 mg, 0.53 mmol) and LiCl (23 mg, 0.53 mmol) was dissolved in 3 mL of CH$_3$CN. After most of the solid had dissolved, DBU (52 mg, 51 µl, 0.34 mmol) was added, changing the colour of the solution to bright yellow. The filtered solution containing the aldehyde was then added and the mixture stirred at r.t. for 3 h.

The mixture was quenched by addition of NaHCO$_3$ and then extracted 4 times with Et$_2$O. The combined organic layers were washed with brine, dried over MgSO$_4$, filtered and concentrated under reduced pressure. Purification by column chromatography (Hex:EtOAc = 20:1 to 7:1) yielded **239** as a slightly yellow oil. (179 mg, 81%)

^{1}H NMR (400 MHz, CDCl$_3$): δ = 6.99 (dd, *J* = 15.5, 5.6 Hz, 1 H), 6.81 (ddd, *J* = 18.5, 10.5, 1.0 Hz, 1 H), 6.42 (m, 2 H), 6.15 (dd, *J* = 11.0, 10.2 Hz, 1 H), 6.12 (dd, *J* = 8.7, 7.7 Hz, 1 H), 6.06 (dd, *J* = 15.5, 1.0 Hz, 1 H), 5.49 (dt, *J* = 10.9, 7.0 Hz, 1 H), 5.06 (d, *J* = 4.8 Hz, 1 H), 4.87 (dd, 7.0, 1.0 Hz, 2 H), 4.35 (dd, *J* = 8.3, 4.8, 1.0 Hz, 1 H), 3.41 (s, 3 H), 3.27 (m, 1 H), 2.26 (dd, *J* = 12.8, 7.0 Hz, 1 H), 1.88 (ddd, *J* = 12.8, 11.0, 4.7 Hz, 1 H), 1.60-1.20 (m, 12 H), 0.95-0.85 (m, 15 H).

^{13}C NMR (100 MHz, CDCl$_3$): δ = 166.4 (C), 147.6 (CH), 141.1 (CH), 139.8 (CH), 139.1 (CH), 135.9 (CH), 123.1 (CH), 121.3 (CH), 105.7 (CH), 85.6 (CH), 83.2 (CH), 61.0 (CH$_2$), 55.4 (CH$_3$), 49.0 (CH), 29.5 (CH$_2$), 27.6 (CH$_2$), 14.1 (CH$_3$), 10.0 (CH$_2$).

R$_f$ (Hex/EtOAc = 3/1): 0.56

Copper-(I)-thiophen-2-carboxylate

A 100 mL single-neck flask was charged with Cu$_2$O (6.98g, 4.88 mmol) and thiophen-2-carboxylic acid (25 g, 19.05 mmol). 75 mL of toluene were added and the mixture was heated to reflux on a Dean-Stark trap overnight. After cooling to ~60 °C, the slurry was filtered under argon. The residue was first washed with 75 mL of MeOH and then repeatedly with freshly destilled, anhydrous diethyl ether until the filtrate was completely colourless (~ 1 L). After washing with hexanes, the redish powder was dried in a stream of argon. (15.47 g, 83 %).

E-1,2-Bis(tributylstannyl)ethene[95a]

A 250 mL 3-neck flask was charged with lithium acetylide ethylendiamine complex (2.54 g, 28.8 mmol) and 60 mL of THF were added. At 0 °C, Bu$_3$SnCl (7.81 g, 24.0 mmol) was added *via* a dropping funnel over 45 min. After removal of the ice-bath, the mixture was stirred at r.t. overnight.

Before quenching, the reaction was cooled again to 0 °C and 10 mL of H$_2$O were added. The reaction mixture was then concentrated under reduced pressure, before it was extracted 3 times with hexanes. The combined organic layers were washed with brine, dried over magnesium sulfate, filtered and concentrated under reduced pressure. Kugelrohr distillation (0.5 mbar, 97 – 100 °C) yielded acetylene-tributyltin (2.65 g, 35 %).

A 25 mL single-neck flask was charged with acetylene-tributyltin ((2.65 g, 8.40 mmol), Bu$_3$SnH (2.93 g, 2.67 mL, 10.08 mmol) and AIBN (0.04 g, 0.21 mmol). The resulting mixture was heated to 90 °C for 1 h. After cooling to r.t. subjection to Kugelrohr distillation (0.05 mbar, 170 °C) yielded pure **243**. (5.36 g, quant.)

^{1}H NMR (250 MHz, CDCl$_3$): δ = 6.89 (s, 2 H), 1.70-1.20 (m, 12 H), 0.89 (t, J = 7.1 Hz, 15 H).

^{13}C NMR (63 MHz, CDCl$_3$): δ = 153.4 (CH), 29.5 (CH$_2$), 27.6 (CH$_2$), 14.1 (CH$_3$), 10.0 (CH$_2$).

(*E*)-3-[(2*S*,3*S*,5*R*)-3-((*Z*)-2-Iodo-vinyl)-5-methoxy-tetrahydro-furan-2-yl]-acrylic acid (*Z*)-3-iodo-allyl ester

Compound **216** (126 mg, 0.44 mmol) was dissolved in 3 mL of CH$_3$CN and IBX (373 mg, 1.33 mmol) was added. The suspension was heated to reflux for 30 min; after cooling to r.t., the solid was filtered to give a solution of **217** in CH$_3$CN.

In a separate flask, compound **247** (245 mg, 0.68 mmol) and LiCl (29 mg, 0.68 mmol) was dissolved in 4 mL of CH$_3$CN. After most of the solid had dissolved, DBU (67 mg, 66 µl, 0.44 mmol) was added, changing the colour of the solution to bright yellow. The filtered solution containing the aldehyde was then added and the mixture stirred at r.t. for 3 h.

The mixture was quenched by addition of NaHCO$_3$ and then extracted 4 times with Et$_2$O. The combined organic layers were washed with brine, dried over MgSO$_4$, filtered and concentrated under reduced pressure. Purification by column chromatography (Hex:EtOAc = 5:1 + 1% NEt$_3$) yielded **244** as a slightly yellow oil. (130 mg, 60%)

^{1}H NMR (250 MHz, CDCl$_3$): δ = 7.01 (dd, J = 15.5, 5.5 Hz, 1 H), 6.53-6.42 (m, 3 H), 6.12 (dd, J = 8.9, 7.8 Hz, 1 H), 6.05 (dd, J = 15.5, 1.5 Hz, 1 H), 5.06 (d, J = 4.8 Hz, 1 H), 4.73 (d, J = 4.5 Hz, 2 H), 4.35 (dd, J = 7.7, 5.8 Hz, 1 H), 3.41 (s, 3 H), 3.27 (m, 1 H), 2.26 (dd, J = 12.6, 7.0 Hz, 1 H), 1.88 (ddd, J = 12.6, 11.2, 4.8 Hz, 1 H).

^{13}C NMR (63 MHz, CDCl$_3$): δ = 166.2 (C), 148.2 (CH), 139.8 (CH), 136.2 (CH), 120.8 (CH), 105.8 (CH), 85.7 (CH), 85.1 (CH), 83.1 (CH), 67.4 (CH$_2$), 55.4 (CH$_3$), 49.1 (CH), 39.4 (CH$_2$).

MS (EI) m/z: 489 (M$^+$, 1), 458 (1), 333 (2), 276 (2), 224 (78), 167 (8), 97 (100).

HRMS(EI) calcd. for C$_{13}$H$_{16}$O$_4$Si$_2$: 489.9138 Found: 489.9145.

R$_f$ (Hex/EtOAc = 3/1): 0.35

<u>O</u>-Acetyl-[(2*R*,3*S*,5*R*)-3-((*Z*)-2-Iodo-vinyl)-5-methoxy-tetrahydro-furan-2-yl]-methanol

Compound **216** (45 mg, 0.16 mmol) was dissolved in 1 mL of pyridine, followed by addition of AcCl (25 mg, 22 μl, 0.32 mmol).

After stirring at r.t. overnight, the reaction was diluted with diethyl ether and quenched by addition of H$_2$O. The mixture was then extracted 2 times with diethyl ether. The combined organic layers were washed with 0.5 M KHSO$_4$ and brine, dried over MgSO$_4$, filtered and concentrated under reduced pressure. Purification by column chromatography (Hex:EtOAc = 5:1 + 1% NEt$_3$) yielded **255** as a colourless oil. (43 mg, 83%)

¹H NMR (250 MHz, CDCl₃): δ = 6.36 (d, *J* = 7.4 Hz, 1 H), 6.07 (dd, *J* = 8.9, 7.4 Hz, 1 H), 5.00 (d, *J* = 4.8 Hz, 1 H), 4.23 (dd, *J* = 10.9, 3.3 Hz, 1 H), 4.07 (dd, *J* = 10.9, 6.6 Hz, 1 H), 4.00 (ddd, *J* = 7.2, 6.6, 3.3 Hz, 1 H), 3.33 (s, 3 H), 3.23 (m, 1 H), 2.21 (dd, *J* = 12.7, 7.2 Hz, 1 H), 2.07 (s, 3 H), 1.81 (ddd, *J* = 12.7, 11.1, 4.8 Hz, 1 H).

¹³C NMR (63 MHz, CDCl₃): δ = 171.1 (C), 140.7 (CH), 140.7 (CH), 105.5 (CH), 84.7 (CH), 82.0 (CH), 66.8 (CH₂), 55.0 (CH₃), 45.7 (CH₃), 39.2 (CH₂), 21.33 (CH).

R$_f$ (Hex/EtOAc = 1/1): 0.52

((2*R*,3*S*,5*R*)-5-methoxy-3-((*1Z*,*3E*)-4-(tributylstannyl)buta-1,3-dienyl)-tetrahydrofuran-2-yl)methanol (Method A)

Compound **338** (43 mg, 0.13 mmol) was dissolved in 5 mL of toluene, which was then subjected to 3 freeze/pump/thaw cycles. Pd(PPh₃)₄ (15 mg, 0.13 mmol) was added, followed by bis-stannane **243** (400 mg, 0.66 mmol).

After stirring at r.t. overnight, the reaction was diluted with diethyl ether and was then washed twice with H₂O, followed by sat. KF solution and brine, dried over MgSO₄, filtered and concentrated under reduced pressure. Purification by column chromatography (Hex:EtOAc = 50:1 to 20:1 + 1% NEt₃) yielded the dienyl stannane acetate as a colourless oil.

The acetate was then dissolved in 5 mL diethyl ether, cooled to 0 °C and DIBAl-H in toluene (0.22 mL, 0.33 mmol, 1.5M) was added dropwise. After stirring at 0 °C for 30 min, the reaction was quenched by addition of 5 mL of KNa-tartrate and the resulting mixture was stirred vigorously at r.t. overnight.

After separation of the phases, the aqueous layer was extracted twice with diethyl ether. The combined organic layers were washed with brine, dried over MgSO$_4$, filtered and concentrated under reduced pressure. Purification by column chromatography (Hex:EtOAc = 20:1 to 5:1 + 3 % NEt$_3$) yielded **256** as a colourless oil. (57 mg, 93 %).

^{1}H NMR (400 MHz, CDCl$_3$): δ= 6.62 (dd, J = 18.6, 10.5 Hz, 1 H), 6.12 (d, J = 18.6 Hz, 1 H), 5.87 (dd, J = 10.5, 10.3 Hz, 1 H), 5.14 (dd, J = 10.9, 10.3 Hz, 1 H), 4.92 (dd, J = 5.5, 3.0 Hz, 1 H), 3.65 (m, 2 H), 3.39 (bd, J = 7.7 Hz, 1 H), 3.22 (s, 3 H), 2.46 (dddd, J = 10.9, 9.5, 7.8, 6.8 Hz, 1 H), 2.28 (ddd, J = 13.6, 9.5, 5.5 Hz, 1 H), 1.57 (ddd, J = 13.6, 7.8, 3.0 Hz, 1 H), 1.33 (m, 6 H), 1.15 (m, 6 H), 0.77 (m, 15 H).

^{13}C NMR (100 MHz): δ = 141.8 (CH), 136.8 (CH), 134.2 (CH), 130.2 (CH), 105.9 (CH), 83.4 (CH), 62.8 (CH$_2$), 55.6 (CH$_3$), 40.9 (CH$_2$), 38.3 (CH), 29.5 (CH$_2$), 27.7 (CH$_2$), 14.1 (CH$_3$), 10.0 (CH$_2$).

IR [cm^{-1}]: 3468 (br), 3070, 2955, 2927, 2871, 2854, 1732, 1694, 1682, 1633, 1589, 1557, 1464, 1428, 1398, 1376, 1339, 1293, 1212, 1110, 1029.

HRMS(EI) calcd. for C$_{18}$H$_{33}$O$_3$Sn (M-Bu): 417.1446, Found: 417.1455.

α$_D^{20}$ - 40.7 (c 1.83, CH$_2$Cl$_2$)

R$_f$ (Hex/EtOAc = 3/1): 0.11

2-((*E*)-3-Tributylstannanyl-allylsulfanyl)-benzothiazole[119]

DIAD (4.57 g, 4.38 mL, 22.6 mmol) was dissolved in 90 mL of THF and cooled to 0 °C. Then a solution of **188** (7.21 g, 20.8 mmol), PPh$_3$ (6.00 g, 22.9 mmol) and **311** (3.68 g, 22.0 mmol) in 50 mL of THF was added *via* a dropping funnel within 10 min. After stirring overnight (and slowly warming to r.t.), the reaction was quenched by addition of 40 mL of sat. NaHCO$_3$ and then filtered over a pad of celite.

The biphasic mixture was then separated and subsequently, the aqueous phase was extracted 4 times with diethyl ether. The combined organic layers were washed with brine, dried over MgSO$_4$, filtered and concentrated under reduced pressure. Purification by column chromatography (Hex:EtOAc = 50:1 + 1% NEt$_3$) yielded **312** as a colourless oil. (10.06 g, 89%)

^{1}H NMR (250 MHz, CDCl$_3$): δ = 7.87 (d, *J* = 8.2 Hz, 1 H), 7.75 (d, *J* = 8.0 Hz, 1 H), 7.41 (t, *J* = 7.7 Hz, 1 H), 7.29 (t, *J* = 7.7 Hz, 1 H), 6.35 (d, *J* = 18.7 Hz, 1 H), 6.10 (dt, *J* = 18.6, 6.2 Hz, 1 H), 4.03 (d, *J* = 6.2 Hz, 2 H), 1.50-1.20 (m, 12 H), 0.88 (t, *J* = 8.0 Hz, 6 H), 0.86 (t, *J* = 7.2 Hz, 1 H).

^{13}C NMR (63 MHz, CDCl$_3$): δ = 141.3 (CH), 135.2 (CH), 126.4 (CH), 124.6 (CH), 122.0 (CH), 121.3 (CH), 40.0 (CH$_2$), 29.4 (CH$_2$), 27.6 (CH$_2$), 14.0 (CH$_3$), 10.0 (CH$_2$).

R$_f$ (Hex/EtOAc = 3/1): 0.76

<u>2-((*E*)-3-Tributylstannanyl-prop-2-ene-1-sulfonyl)-benzothiazole</u>[119]

(NH$_4$)$_6$Mo$_7$O$_{24}$·7H$_2$O (5.01 g, 4.0 mmol) was dissolved in 30 % H$_2$O$_2$ (23 mL, 202.6 mmol) to give a bright-yellow solution. After cooling to 0 °C, a solution of **312** (10.1 g, 20.3 mmol) in 35 mL of EtOH was added *via* dropping funnel within 20 min. After stirring at 0 °C for 24 h, reaction control *via* TLC showed no remaining starting material.

120 mL of brine were added and the mixture was then extracted 4 times with diethyl ether. The combined organic layers were washed with brine, dried over MgSO$_4$, filtered, concentrated under reduced pressure, and the residue was treated twice with toluene, which was also removed under reduced pressure.

The crude product was recrystallised from MeOH to give **313** as waxy, yellow crystals (10.61 g, 99 %)

^{1}H NMR (250 MHz, CDCl$_3$): δ = 8.22 (m, 1 H), 7.98 (m, 1 H), 7.61 (m, 2 H), 6.32 (dt, *J* = 18.7, 1.1 Hz, 1 H), 5.95 (dt, *J* = 18.7, 6.8 Hz, 1 H), 4.28 (dd, *J* = 6.8, 1.1 Hz, 1 H), 1.40-1.10 (m, 12 H), 0.95-0.64 (m, 15 H).

^{13}C NMR (63 MHz, CDCl$_3$): δ = 145.6 (CH), 131.5 (CH), 128.3 (CH), 128.0 (CH), 127.9 (CH), 125.9 (CH), 122.6 (CH), 62.7 (CH$_2$), 29.2 (CH$_2$), 27.5 (CH$_2$), 14.0 (CH$_3$), 10.0 (CH$_2$).

R$_f$ (Hex/EtOAc = 3/1): 0.60

tert-Butyl-[(*2R,3S,5R*)-5-methoxy-3-((*1Z,3E*)-4-tributylstannanyl-buta-1,3-dienyl)-tetrahydro-furan-2-ylmethoxy]-diphenyl-silane

Aldehyde **207a** (4.00 g, 10.04 mmol) and sulfone **313** (7.92g, 15.0 mmol) were dissolved in 300 mL toluene and cooled to -78 °C, then KHMDS in toluene (30.0 mL,

15.0 mmol, 0.5 M) was added slowly within one hour. The reaction continued to stir for 8 hours at -78 °C and then warmed to room temperature overnight.

The reaction was quenched by addition of 100 mL of sat. NaHCO$_3$ solution. After separation of the organic layer, the aqueous layer was extracted with a total of 200 mL of Et$_2$O. The combined organic layers were washed with brine, dried over magnesium sulfate, filtered and the solvent removed under reduced pressure. Purification by column chromatography using Hex:EtOAc = 20:1 to 7:1 yielded **265** (4.36 g, 61%) as a colourless oil.

^{1}H NMR (400 MHz, CDCl$_3$): δ = 7.61 (m, 4 H), 7.30 (m, 6 H), 6.67 (dd, *J* = 18.6, 10.4 Hz, 1 H), 6.18 (d, *J* = 18.6 Hz, 1 H), 5.89 (dd, *J* = 10.7, 10.4 Hz, 1 H), 5.23 (dd, *J* = 10.7, 10.7 Hz, 1 H), 5.00 (dd, *J* = 5.3, 2.7 Hz, 1 H), 3.79 (m, 1 H), 3.72 (dd, *J* = 11.1, 3.1 Hz, 1 H), 3.63 (dd, *J* = 11.1, 4.8 Hz, 1 H), 3.31 (s, 3 H), 3.14 (dddd, *J* = 10.7, 9.5, 7.1, 6.2 Hz, 1 H), 2.35 (ddd, *J* = 13.4, 9.5, 5.3 Hz, 1 H), 1.61 (ddd, *J* = 13.4, 7.1, 2.7 Hz, 1 H), 1.5 (m, 6 H), 1.28 (m, 6 H), 1.06 (s, 9 H), 0.88 (m, 15 H).

^{13}C NMR (100 MHz, CDCl$_3$): δ = 141.5 (CH), 135.7 (CH), 135.6 (CH), 133.6 (C), 132.9 (CH), 130.8 (CH), 129.5 (CH), 127.6 (CH), 105.3 (CH), 83.9, (CH), 64.7 (CH$_2$), 55.0 (CH$_3$), 40.4 (CH$_2$), 38.5 (CH), 29.1 (CH$_2$), 27.2 (CH$_2$), 26.8 (CH$_3$), 19.3 (C), 13.7 (CH$_3$), 9.5 (CH$_2$).

HRMS(EI) calcd. For C$_{34}$H$_{51}$O$_3$SiSn (M-Bu): 655.2624, Found: 655.2638.

R$_f$ (Hex/EtOAc = 3/1): 0.66

((2R,3S,5R)-5-methoxy-3-((1Z,3E)-4-(tributylstannyl)buta-1,3-dienyl)-tetrahydrofuran-2-yl)methanol

To a solution of compound **265** (0.75g, 1.06 mmol) in 10 mL THF, TBAF in THF (1.17 mL, 1.17 mmol, 1 M) was added. After 2 hours, the reaction was quenched by addition of saturated NaHCO$_3$ solution and subsequently extracted 4 times with a total of 100 mL of ethyl acetate. The combined organic layers were washed with brine, dried over magnesium sulfate, filtered and the solvent removed under reduced pressure. Purification by column chromatography using Hex:EtOAc = 20:1 to 5:1 (+3% NEt$_3$) yielded **256** (0.43 g, 85%) as a colourless oil.

Spectroscopic analysis was identical to that of **256** obtained by Method A

Methyl *(Z)*-3-Iodo-propenoate

A 100 mL single-neck flask was charged with methyl propiolate (**245**) (5.0 g, 59.47 mmol) and dissolved in 50 mL CH$_3$CN. Then, lithium iodide (8.76g, 65.41 mmol) and acetic acid (3.93 mL, 65.41 mmol) were added. The resulting solution was heated under vigorous stirring to reflux. After 10 min, a voluminous white precipitate formed. Stirring was continued overnight at reflux. After cooling, the reaction mixture was neutralised by pouring into 100 mL of 0.3M K$_2$CO$_3$ solution. The resulting solution was extracted 4 times with a total of 200 mL diethyl ether. The combined organic layers were washed with brine, dried over MgSO$_4$. Removal of the solvents under reduced pressure yielded the crude enoate, which was directly used without any further purification.

^{1}H NMR (250 MHz, CDCl$_3$): δ = 7.46 (d, J = 8.9 Hz, 1 H), 6.90 (d, J = 8.9 Hz, 1 H), 3.78 (s, 3 H). *cf.* Ref. 96

A 1 L three-necked flask with dropping funnel and thermometer was charged with (Z)-3-iodopropenoate (10.96 g, 51.70 mmol) dissolved in 250 mL diethyl ether. The solution was cooled to 0 °C and DIBAl-H in toluene (72 mL, 108 mmol, 1.5 M) was added slowly. 30 minutes after the of addition, reaction control by TLC showed full consumption of the starting material. The reaction was hydrolysed by addition of 10 mL MeOH. Subsequently, 250 mL saturated KNa-tartrate solution were added and the reaction was stirred vigorously, until two clear layers formed (~ 1 hour).

The two layers were separated, and the aqueous layer was extracted twice with 100 mL of diethyl ether. The combined organic layers were washed with brine, dried over MgSO$_4$ and the solvents removed under reduced pressure. Kugelrohr-destillation under high vacuum yielded **246** as a slightly yellow oil which was stored at -20 °C. (8.28 g, 87%)

^{1}H NMR (250 MHz, CDCl$_3$): δ = 6.47 (dt, J = 7.6, 5.6 Hz, 1 H), 6.34 (dt, J = 7.6, 1.4 Hz, 1 H), 4.21 (bm, 1 H), 2.32 (bs, 1 H).

^{13}C NMR: (63 MHz, CDCl$_3$): δ = 140.4 (CH), 83.0 (CH), 66.1 (CH$_2$).

R$_f$ (Hex/EtOAc = 1/1): 0.44

(Z)-3-iodoallyl 2-(diethoxyphosphoryl)acetate

Phosponoacetic acid (**233**) (3.31 g, 16.86 mmol) was dissolved in 40 mL of DCM and cooled to 0 °C. Oxalyl chloride (2.21 mL, 25.29 mmol) was added slowly. After stirring at 0° C for 20 min, the reaction mixture was warmed to room temperature and stirring continued for 2 h. Then the volatile components were removed under reduced pressure and the acid chloride **234** was used without any further purification.

Compound **246** (1.55 g, 8.43 mmol) was dissolved in 60 mL DCM and cooled to 0 °C. Pyridine (2.05 mL, 25.59 mmol) and DMAP (102 mg, 0.82 mmol) were added, then the acid chloride (dissolved in 20 mL DCM) was added slowly *via* cannula. The reaction was warmed to room temperature and stirred overnight.

The reaction was quenched by addition of 0.5 M KHSO₄ solution. After separation of the organic layer, the aqueous layer was subsequently extracted 4 times with a total of 200 mL of DCM. The combined organic layers were washed with brine, dried over magnesium sulfate, filtered and concentrated under reduced pressure. Purification by column chromatography (Hex:ethylacetate = 1:2) yielded **247** as a slightly yellow oil (2.66 g, 87%).

¹H NMR (250 MHz, CDCl₃): δ = 6.52 (d, J = 7.9 Hz, 1 H), 6.45 (dt, J = 7.9, 5.4 Hz, 1 H), 4.72 (d, J = 5.4 Hz, 2 H), 4.17 (dq, J = 7.9, 7.1 Hz, 2 H), 2.99 (d, J = 21.6 Hz, 2 H), 1.34 (t, J = 7.1 Hz, 3 H).

¹³C NMR (75 MHz, CDCl₃): δ = 165.4 (d, J = 6.4 Hz, C), 135.0 (CH), 85.1 (CH), 67.8 (CH₂), 62.4 (d, J = 6.4 Hz, CH₂), 34.1 (d, J = 133.4 Hz, CH₂), 16.2 (d, J = 6.0 Hz, CH₃).

R_f (Hex/EtOAc = 1/1): 0.11

(*E*)-((*Z*)-3-iodoallyl) 3-((2*S*,3*S*,5*R*)-5-methoxy-3-(((1*Z*,3*E*)-4-(tributylstannyl)buta-1,3-dienyl)-tetrahydrofuran-2-yl)acrylate

A 25 mL single-neck flask was charged with activated 4Å molecular sieves (80 mg). After addition of 2 mL of DCM, **256** (190 mg, 0.40 mmol) and NMO (49 mg, 0.42 mmol) was added and the suspension stirred for 15 min. Then, TPAP (7 mg, 0.02 mmol) was added and the reaction closely monitored *via* TLC (completion usually within 30 minutes).
In a separate 25 mL single-neck flask, LiCl (dried for 2 h in vacuo at 120 °C, 26 mg, 0.60 mmol) was suspended in 6 mL CH₃CN. Then phosphonate **247** (218 mg, 0.60

mmol) and DBU (80 mg, 0.52 mmol) was added and the resulting yellow solution was stirred for 5 min. Then it was transferred via cannula to the flask with the TPAP oxidation.

After stirring for one hour, the reaction was quenched by addition of 10 mL of sat. $NaHCO_3$. After extracting 4 times with a total of 50 mL of Et_2O, the combined organic layers were washed with brine, dried over magnesium sulfate, filtered and the solvent removed under reduced pressure. Purification by column chromatography (Hex:EtOAc = 20:1 (+ 1 % NEt_3)) yielded **251** (204 mg, 75 %) as a colourless oil.

^{1}H NMR (250 MHz, $CDCl_3$): δ = 6.95 (dd, J = 15.7, 5.0 Hz, 1 H), 6.67 (dd, J = 18.5, 10.4 Hz, 1 H), 6.49 (d, J = 7.8 Hz, 1 H), 6.43 (dt, J = 7.8, 4.0 Hz, 1 H), 6.30 (d, J =18.5 Hz, 1 H), 6.09 (d, J = 15.7 Hz, 1 H), 6.06 (dd, J = 10.4, 10.3 Hz, 1 H), 5.31 (dd, J = 11.0, 10.3 Hz, 1 H), 5.13 (dd, J = 5.5, 3.0 Hz, 1 H), 4.71 (d, J = 4.0 Hz, 2 H), 4.32 (dd, J = 8.6, 5.0 Hz, 1 H), 3.39 (s, 3 H), 3.05 (dddd, J = 11.0, 9.4, 8.6, 7.8 Hz, 1 H), 2.49 (ddd, J = 13.6, 9.4, 5.5 Hz, 1 H), 1.75 (ddd, J = 13.6, 7.8, 3.0 Hz, 1 H), 1.5 (m, 6 H), 1.3 (m, 6 H), 0.88 (m, 15 H).

^{13}C NMR (75 MHz, $CDCl_3$): 169.5 (C), 146.5 (CH), 141.5 (CH), 137.6 (CH), 136.2 (CH), 134.6 (CH), 128.7 (CH), 120.9 (CH), 105.8 (CH), 84.9 (CH), 81.6 (CH), 67.4 (CH_2), 55.7 (CH_3), 43.9 (CH), 40.8 (CH_2), 29.5 (CH_2), 27.6 (CH_2), 14.1 (CH_3).

IR [cm^{-1}]: 2955, 2926, 2871, 2852, 1729, 1699, 1661, 1651, 1434, 1455, 1376, 1338, 1303, 1273, 1213, 1160, 1106, 1027.

MS (EI) m/z 623 (M$^+$-Bu, 4), 361 (4), 279 (30), 149 (100).

HRMS(EI) calcd. For $C_{23}H_{36}IO_4Sn$ (M-Bu): 623.0685, Found: 623.0698.

$α_D^{20}$ +38 (c 1.2, CH_2Cl_2)

R_f (Hex/EtOAc = 3/1): 0.63

(*E*)-3-[(2*S*,3*S*,5*R*)-3-((*Z*)-But-1-en-3-ynyl)-5-methoxy-tetrahydro-furan-2-yl]-acrylic acid (*Z*)-3-iodo-allyl ester

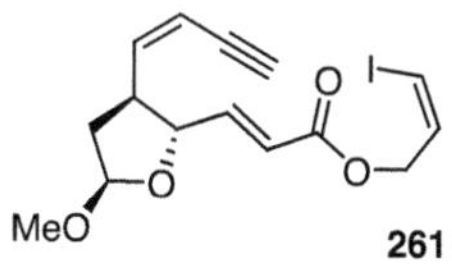

¹H NMR (250 MHz, CDCl₃): δ = 7.01 (dd, *J* = 15.5, 5.5 Hz, 1 H), 6.51 (d, *J* = 7.3 Hz, 1 H), 6.48 (dd, *J* = 5.3, 2.5 Hz, 1 H), 6.06 (dd, *J* = 15.5, 1.6 Hz, 1 H), 5.84 (dd, *J* = 10.7, 10.3 Hz, 1 H), 5.59 (dd, *J* = 10.7, 2.3 Hz, 1 H), 5.06 (d, *J* = 4.8 Hz, 1 H), 4.73 (m, 2 H), 4.31 (ddd, *J* = 8.8, 5.3, 0.6 Hz, 1 H), 3.53 (dddd, *J* =, 11.0, 9.1, 9.0, 7.0 Hz, 1 H), 3.41 (s, 3 H), 2.23 (dd, *J* = 12.7, 7.0 Hz, 1 H), 1.89 (ddd, *J* = 12.7, 11.4, 5.0 Hz, 1 H).

¹³C NMR (63 MHz, CDCl₃): δ = 146.7 (CH), 144.4 (CH), 136.1 (CH), 120.8 (CH), 110.9 (CH), 105.9 (CH), 85.1 (CH), 83.0 (C), 81.3 (CH), 67.4 (CH₂), 55.6 (CH₃), 45.5 (CH), 39.5 (CH₂).

R$_f$ (Hex/EtOAc = 3/1): 0.53

(2R,3aS,4Z,6E,8Z,13E,14aS)-2-methoxy-3,3a-dihydrofuro[2,3-*e*][1]oxacyclotridecin-12(2H,10H,14aH)-one (Method A):

Seco-compound **251** was dissolved in 20 mL DMF in a Schlenk flask and then degassed in 2. Then Pd$_2$dba$_3$·CHCl$_3$ (7 mg, 0.01 mmol), P(o-furyl)$_3$ (3 mg, 0.02 mmol) and CuI (15 mg, 0.08 mmol) were added sequentially. The resulting mixture turned brown within minutes and was stirred for 36 h at room temperature.

The solvent was removed under vacuum and the residue redissolved in ethyl acetate. Saturated NaHCO$_3$ solution was added, and, after separation of the organic layer, was subsequently extracted twice with ethyl acetate. The combined organic layers were washed with brine, dried over MgSO$_4$, and then concentrated under reduced pressure. Purification by column chromatography (Hex:EtOAc = 10:1) yielded **224** as a colourless oil. (9 mg, 51 %)

^{1}H NMR (250 MHz, CDCl$_3$): δ = 6.85 (dd, *J* =16.0, 6.8 Hz, 1 H), 6.26-5.95 (m, 4 H), 5.87 (d, *J* = 16.1 Hz, 1 H), 5.59 (m, 1 H), 5.35 (dd, *J* = 10.2 Hz, 1 H), 5.14 (dd, *J* = 5.7, 3.9 Hz, 1 H), 4.92 (dd, *J* = 15.0, 5.3 Hz, 1 H), 4.72 (bd, *J* = 15.0 Hz, 1 H), 4.18 (dd, *J* = 9.7, 9.0 Hz, 1 H), 3.40 (s, 3 H), 2.75 (m, 1 H), 2.52 (ddd, *J* = 13.5, 9.0, 5.7 Hz, 1 H), 1.76 (ddd, *J* =13.5, 9.7, 3.9 Hz, 1 H).

^{13}C NMR (75 MHz, CDCl$_3$): δ = 169.2 (C), 147.1 (CH), 131.8 (CH), 130.4 (CH), 129.1 (CH), 128.9 (CH), 128.8 (CH), 126.6 (CH), 120.9 (CH), 106.1 (CH), 80.7 (CH), 61.6 (CH$_2$), 55.6 (CH$_3$), 46.9 (CH), 40.0 (CH$_2$).

IR [cm^{-1}]: 2930, 1725, 1654, 1622, 1577, 1496, 1449, 1372, 1340, 1243, 1214, 1154, 1099, 1028, 981, 952, 914, 858, 768, 744, 670.

MS (EI) m/z 262 (M$^+$, 45), 231 (6), 183 (29), 162 (91).

HRMS (EI) calcd. For $C_{15}H_{18}O$: 262.1205, Found: 262.1212.

α_D^{20} -57.0 (c 1.33, CH_2Cl_2)

R_f (Hex/EtOAc = 3/1): 0.53

Tetrabutyl ammonium diphenylphospinate

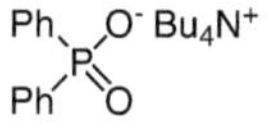

diphenylphosphinic acid (2.0 g, 9.2 mmol) was dissolved in 10 mL of MeOH tetra-*n*-butylammonium hydroxide in MeOH (9.2 mL, 1 M) was added and the mixture stirred briefly. After filtration through a pad of celite, the solvent was removed under reduced pressure to yield a yellow oil, which was recrystallized from diethyl ether to give colourless, hygroscopic needles. (3.76 g, 91 %)

(2R,3aS,4Z,6E,8Z,13E,14aS)-2-methoxy-3,3a-dihydrofuro[2,3-e][1]oxacyclotridecin-12(2H,10H,14aH)-one (5) (Method B):

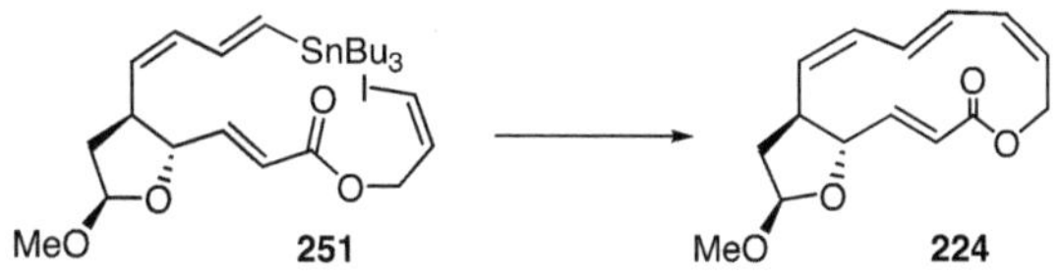

A 250 mL Schlenk-flask was charged with freshly distilled THF (140 mL). The solvent was degassed in 3 freeze/pump/thaw cycles. From this volume, 3 mL THF were removed to dissolve *seco*-compound **251** (100 mg, 0.15 mmol). To the main volume Pd(PPh$_3$)$_4$ (34 mg, 0.03 mmol) and LiCl (9.4 mg, 0.22 mmol) was added, followed by

251 dissolved in THF. After 5 min, Bu$_4$NPh$_2$PO$_2$ (135 mg, 0.29 mmol) was added. The solution was then heated to 40 °C for 3 h, after which another 10 mg of Pd(PPh$_3$)$_4$ and 4 mg LiCl were added, and the reaction stirred at 40 °C for 18 h.

The reaction mixture was filtered over a pad of celite and the solvent removed under reduced pressure. The residue was redissolved in hexanes, and washed with saturated NaHCO$_3$-solution. After separation of the organic layer, the aqueous layer was extracted 3 times with ethyl acetate. The combined organic layers were washed with brine, dried over MgSO$_4$ and then concentrated under reduced pressure. Purification by column chromatography (Hex:EtOAc = 10:1 to 7:1) yielded **224** as a colourless oil. (25 mg, 65%)

Spectroscopic analysis was identical to that of **224** obtained by Method A

(2R,3aR,5aR,7aS,10aS,10bS,10cR)-2-methoxy-2,3,3a,7a,8,10a,10b,10c-octahydro-5a*H*-1,9-dioxa-dicyclopenta[a,h]naphthalen-10-one

Macrolactone **224** (20 mg, 0.07 mmol) and BHT (5 mg, 0.02 mmol) were dissolved in 3 mL of xylenes and heated to reflux for 24 h. The reaction mixture was concentrated under reduced pressure and purified by column chromatography (Hex:EtOAc = 2:1) to yield **225** as a white solid. (14 mg, 70 %)
m.p. 114-118 °C.

^{1}H NMR (600 MHz, CDCl$_3$) δ = 5.76 (bd, *J* = 9.6 Hz, 1 H), 5.63 (bd, *J* = 10.2 Hz, 1 H), 5.54 (bd, *J* = 9.8 Hz, 1 H), 5.52 (ddd, *J* = 9.8, 3.6, 3.6 Hz, 1 H) 5.15 (dd, *J* = 5.2 Hz, 1 H), 4.38 (dd, *J* = 8.8, 5.7 Hz, 1 H), 4.14 (d, *J* = 8.8 Hz, 1 H), 3.51 (dd, *J* = 11.7, 9.6 Hz, 1 H), 3.40 (s, 3 H), 3.22 (*J* = 6.9, 2.4 Hz, 1 H), 3.18 (m, 1 H), 3.06 (m, 1 H), 2.73 (ddd,

J =11.6, 6.8, 2.5 Hz, 1 H), 2.44 (J =12.5, 7.4, 5.5 Hz, 1 H), 2.35 (m, 1 H), 1.51 (ddd, J = 12.5, 12.5, 4.9 Hz, 1 H).

^{13}C NMR (125 MHz, CDCl$_3$) δ = 172.9 (C), 131.9 (CH), 130.1 (CH), 125.3 (CH), 123.6 (CH), 106.6 (CH), 77.5 (CH), 72.1 (CH$_2$), 55.8 (CH$_3$), 44.1 (CH), 38.9 (CH), 37.3 (CH$_2$), 35.8 (CH), 34.9 (CH), 32.7 (CH).

IR [cm^{-1}]: 3021, 2917, 2849, 1771, 1546, 1454, 1369, 1211, 1162, 1107, 1088, 1070, 1030, 1000, 967, 937, 906, 858, 757, 728, 680, 517.

MS (EI) m/z 262 (M$^+$, 61), 230 (68), 204, (100).

HRMS(EI) calcd. For C$_{15}$H$_{18}$O: 262.1205, Found: 262.1210.

[α]$_D$ ^{20} -24.6 (c 1.64, CH$_2$Cl$_2$)

R$_f$ (Hex/EtOAc = 1/1): 0.36

(2R,3aS,5aS,7aR,10aR,10bR,10cR)-2-methoxy-2,3,3a,7a,8,10a,10b,10c-octahydro-5a*H*-1,9-dioxa-dicyclopenta[*a,h*]naphthalen-10-one

226

^{1}H NMR (600 MHz, CDCl$_3$) δ = 5.91 (ddd, J = 9.9, 5.6, 2.4 Hz, 1 H), 5.79 (ddd, J = 9.6, 1.9, 1.9 Hz, 1 H), 5.65 (ddd, J = 9.9, 1.7, 1.7 Hz, 1 H), 5.53 (ddd, J = 9.6, 3.9, 2.8 Hz, 1 H), 5.14 (dd, J = 5.3, 4.8 Hz, 1 H), 4.34 (dd, J = 8.7, 5.7 Hz, 1 H), 4.12 (dd, J = 9.1, 2.7 Hz, 1 H), 3.85 (dd, J = 10.1, 6.0 Hz, 1 H), 3.42 (s, 3 H), 3.17 (m, 1 H), 3.02 (ddd, J = 7.4, 6.0, 4.2 Hz, 1 H), 2.95 (dd, J = 7.0, 4.2 Hz, 1 H), 2.92 (m, 1 H), 2.45 (dd, J = 12.7, 7.4, 5.4 Hz, 1 H), 2.30 (m, 1 H), 1.52 (dd, J = 12.7, 5.0 Hz, 1 H).

R_f (Hex/EtOAc = 1/1): 0.33

(Diethoxy-phosphoryl)-acetic acid (Z)-3-trimethylstannanyl-allyl ester

A 100 mL single-neck flask was charged with compound **246** (1.0 g, 5.44 mmol) dissolved in 55 mL of diethyl ether. After cooling to -40 °C, MeLi in diethyl ether (6.52 mL, 6.52 mmol, 1.0 M) was added portionwise within 10 min. The reaction was then warmed to r.t. for 10 min, and then cooled to -78 °C when *n*-BuLi in hexanes (5.0 mL, 12.50 mmol, 2.5 M) was added. After stirring at this temperature for 1.5 h, a solution of Me₃SnCl (3.90 g, 19.57 mmol) in 2 mL of THF was added and the reaction mixture was warmed to r.t. overnight.

The reaction was then quenched by addition of 30 mL of H_2O. After separation of the organic layer, the aqueous layer was then extracted 3 times with diethyl ether. The combined organic layers were washed with brine, dried over $MgSO_4$, filtered and concentrated under reduced pressure to yield a colourless oil which was employed without further purification for the next step. (723 mg, 60%)

¹H NMR (250 MHz, CDCl₃): δ = 6.64 (dt, *J* = 13.0, 5.3 Hz, 1 H), 6.08 (dt, *J* = 13.0, 1.4 Hz, 1 H), 4.18 (ddd, *J* = 5.5, 5.3, 1.4 Hz, 2 H), 1.34 (t, *J* = 5.5 Hz, OH), 0.17 (s, 9 H).

Compound **271** (723 mg, 3.27 mmol) was dissolved in 15 mL of DCM, then **233** (642 mg, 3.27 mmol), EDC·HCl (941 mg, 4.91 mmol) and DMAP (800 mg, 6.55 mmol) were added and the reaction stirred at r.t. for 72 h.

The reaction was then quenched by addition of H_2O, and then extracted 3 times with DCM. The combined organic layers were washed with brine, dried over $MgSO_4$, filtered and concentrated under reduced pressure. Purification by column chromatography (Hex:EtOAc = 1:1 to 1:2 + 5% NEt₃) yielded **272** as a colourless oil. (159 mg, 12 %)

^{1}H NMR (250 MHz, CDCl$_3$): δ = 6.54 (dt, J = 13.0, 6.5 Hz, 1 H), 6.23 (dt, J = 13.0, 1.0 Hz, 1 H), 4.59 (dd, J = 6.5, 1.0 Hz, 2 H), 4.16 (qd, J = 8.3, 7.1 Hz, 4 H), 2.97 (d, J = 21.5 Hz, 2 H), 1.33 (t, J = 7.1 Hz, 6 H), 0.20 (s, 9 H).

^{13}C NMR (63 MHz, CDCl$_3$): δ = 140.9 (CH), 136.9 (CH), 67.7 (CH$_2$), 62.7 (d, J = 6.2 Hz, CH$_2$), 34.32 (d, J = 134.5 Hz, CH$_2$), 16.3 (d, J = 6.2 Hz, CH$_3$), -8.6 (CH$_3$).

MS (EI) *m/z*: 385 (M-Me$^+$, 100), 345 (10), 299 (34), 269 (16), 241 (7), 207 (35), 179 (47).

HRMS(EI) calcd. For C$_{11}$H$_{22}$O$_5$SnP: 385.0229, Found: 385.0235.

[(2R,3S,5R)-3-((1Z,3E)-4-Iodo-buta-1,3-dienyl)-5-methoxy-tetrahydro-furan-2-yl]-methanol

Compound **265** (899 mg, 1.26 mmol) was dissolved in 20 mL of DCM and cooled to -78 °C. A solution of I$_2$ (484 mg, 1.91 mmol) in 10 mL of DCM was added dropwise and the reaction then warmed slowly to r.t.

The reaction was then quenched with 1 M Na$_2$S$_2$O$_3$ and stirred for 30 min. The mixture was extracted 3 times with pentane. The combined organic layers were washed with brine, dried over MgSO$_4$, filtered and concentrated under reduced pressure to yield a colourless oil which was employed without further purification for the next step.

Compound **267** (691 mg, 1.26 mmol) was dissolved in 5 mL of THF and a solution of TBAF in THF (2.52 mL, 2.52 mmol, 1.0 M) was added. After stirring at r.t. overnight, the reaction was quenched by addition of sat. NaHCO$_3$. The mixture was then extracted 4 times with diethyl ether. The combined organic layers were washed with brine, dried over MgSO$_4$, filtered and concentrated under reduced pressure. Purification by column chromatography (Hex:EtOAc = 10:1 to 2:1 + 5 % NEt$_3$) yielded **277** as a colourless oil. (283 mg, 75%)

^{1}H NMR (250 MHz, CDCl$_3$): δ = 7.29 (ddd, J = 14.2, 11.3, 1.1 Hz, 1 H), 6.38 (d, J = 14.2 Hz, 1 H), 5.96 (dd, J = 11.3, 10.8 Hz, 1 H), 5.39 (dd, J = 10.8, 10.4 Hz, 1 H), 5.07 (dd, J = 5.4, 2.8 Hz, 1 H), 3.81 (m, 2 H), 3.53 (dd, J = 12.6, 4.7 Hz, 1 H), 3.37 (s, 3 H), 3.13 (m, 1 H), 2.41 (ddd, J = 13.5, 9.7, 5.4 Hz, 1 H), 1.71 (ddd, J = 13.5, 7.4, 2.8 Hz, 1 H).

^{13}C NMR (63 MHz, CDCl$_3$): δ = 140.4 (CH), 132.2 (CH), 130.2 (CH), 105.4 (CH), 82.5 (CH), 81.1 (CH), 62.1 (CH$_2$), 55.2 (CH$_3$), 40.3 (CH$_2$), 37.8 (CH).

R$_f$ (Hex/EtOAc = 1/1): 0.37

(E)-3-[(2S,3S,5R)-3-((1Z,3E)-4-Iodo-buta-1,3-dienyl)-5-methoxy-tetrahydro-furan-2-yl]-acrylic acid (Z)-3-trimethylstannanyl-allyl ester

Compound **277** (71 mg, 0.23 mmol) was dissolved in 3 mL of CH$_3$CN and IBX (193 mg, 0.69 mmol) was added. The suspension was heated to reflux for 30 min; after cooling to r.t., the solid was filtered to give a solution of **278** in CH$_3$CN.

In a separate flask, compound **272** (104 mg, 0.26 mmol) and LiCl (70 mg, 0.26 mmol) was dissolved in 4 mL of CH$_3$CN. After most of the solid had dissolved, DBU (37 mg, 36 µl, 0.24 mmol) was added, changing the colour of the solution to bright yellow. The filtered solution of **278** was then added and the mixture stirred at r.t. for 3 h.

The mixture was quenched by addition of sat. NaHCO$_3$ and then extracted 4 times with Et$_2$O. The combined organic layers were washed with brine, dried over MgSO$_4$, filtered and concentrated under reduced pressure. Purification by column chromatography (Hex:EtOAc = 5:1 + 1% NEt$_3$) yielded **253** as a colourless oil. (47 mg, 37%)

^{1}H NMR (250 MHz, CDCl$_3$): δ = 7.16 (ddd, J = 14.2, 11.3, 1.1 Hz, 1 H), 6.89 (dd, J = 15.6, 5.3 Hz, 1 H), 6.59 (dt, J = 12.8, 6.4 Hz, 1 H), 6.40 (d, J = 14.2 Hz, 1 H), 6.23 (dt, J = 12.9, 0.9 Hz, 1 H), 6.06 (dd, J = 15.6, 1.6 Hz, 1 H), 6.00 (dd, J = 11.3, 10.7 Hz, 1 H), 5.41 (dd, J = 10.7, 10.3 Hz, 1 H), 5.12 (dd, J = 5.5, 2.8 Hz, 1 H), 4.62 (dd, J = 6.4, 0.9 Hz, 2 H), 4.31 (ddd, J = 8.6, 5.3, 1.6 Hz, 1 H), 3.38 (s, 3 H), 2.93 (m, 1 H), 2.46 (ddd, J = 13.6, 9.5, 5.5 Hz, 1 H), 1.73 (ddd, J = 13.6, 7.5, 2.8 Hz, 1 H), 0.21 (s, 9 H).

^{13}C NMR (63 MHz, CDCl$_3$): δ = 165.3 (C), 145.2 (CH), 141.4 (CH), 140.2 (CH), 136.2 (CH), 130.7 (CH), 130.6 (CH), 121.3 (CH), 105.3 (CH), 81.7 (CH), 80.9 (CH), 66.7 (CH$_2$), 55.3 (CH$_2$), 43.6 (CH), 40.0 (CH$_2$), -8.5 (CH$_3$).

R_f (Hex/EtOAc = 3/1): 0.45

(1-Bromo-allyl)-trimethyl-silane[108]

Diisopropylamine (3.33 g, 4.66 mL, 33.0 mmol) was dissolved in 20 mL of THF and cooled to 0 °C. n-BuLi in hexanes (12mL, 30 mmol, 2.5 M) was added and, after stirring at 0 °C for 30 min, the reaction mixture was cooled to -60 °C. Then, a solution of allyl bromide (7.26 g, 5.2 mL, 60 mmol) and TMSCl (4.89 g, 5.75 mL, 45 mmol) in 20 mL of THF was added within 10 min. After stirring at 30 min, 2 mL of H_2O were added and the mixture was warmed to r.t.

Another 20 mL of 1M HCl were added and the mixture was then extracted 3 times with pentane. The combined organic layers were washed with brine and dried over $MgSO_4$. The solvent was removed by careful distillation over a Vigreux-column, followed by fractional distillation under reduced pressure (~20 mbar, 47-52 °C), yielded **286** as a colourless liquid. (8.11g, 70 %)

^1H NMR (250 MHz, $CDCl_3$): δ = 5.92 (ddd, J = 16.7, 10.2 Hz, 1 H), 5.17 (d, J = 16.7 Hz, 1 H), 5.04 (d, J = 10.2 Hz, 1 H), 3.80 (d, J = 9.6 Hz, 1 H), 0.14 (s, 9 H). *cf.* Ref. 108

[(2R,3S,5R)-3-((Z)-Buta-1,3-dienyl)-5-methoxy-tetrahydro-furan-2-yl]-methanol

Vacuum-dried CrCl$_3$ (244 mg, 1.54 mmol) was suspended in 1 mL of THF. After cooling the suspension to 0 °C, a solution of LiAlH$_4$ in THF (0.61 mL, 0.01 mmol, 1 M) was added slowly. After stirring at 0 °C for 20 min, to the tarry-brown suspension a solution of **207a** (110 mg, 0.28 mmol) and **286** (160 mg, 0.83 mmol) in 1.5 mL of THF was added. After stirring at 0 °C for 45 min, the reaction was allowed to warm to r.t. and stirred overnight.

The violet-brownish suspension was then quenched by addition of pH 7-buffer and extracted 4 times with EtOAc. The combined organic layers were washed with brine, dried over MgSO$_4$, filtered and concentrated under reduced pressure.

The crude produced was redissolved in 1 mL of THF and added at 0 °C to a suspension of KH (100 mg, 1.66 mmol, 35%) in 1 mL of THF. After stirring for 15 min, the reaction was quenched by pouring onto ice/water. The mixture was then extracted 3 times with diethyl ether. The combined organic layers were washed with brine, dried over MgSO$_4$, filtered and concentrated under reduced pressure. Purification by column chromatography (Hex:EtOAc = 5:1 to 2:1 + 2 % NEt$_3$) yielded **289** as a colourless oil. (14 mg, 27 %)

^{1}H NMR (400 MHz, CDCl$_3$): δ = 6.66 (dddd, J = 16.7, 10.9, 10.3 Hz, 1 H), 6.09 (dd, J = 11.0, 10.6 Hz, 1 H), 5.29 (dd, J = 10.5, 10.2 Hz, 1 H), 5.24 (d, J = 16.7 Hz, 1 H), 5.16 (d, J = 10.3 Hz, 1 H), 5.00 (d, J = 5.0 Hz, 1 H), 3.89 (ddd, J = 8.3, 3.6, 3.0 Hz, 1 H), 3.76 (bd, J = 11.6 Hz, 1 H), 3.51 (m, 1 H), 3.40 (s, 3 H), 2.20 (s, OH), 2.16 (dd, J = 12.9, 7.3 Hz, 1 H), 1.84 (ddd, J = 12.9, 11.2, 5.0 Hz, 1 H).

^{13}C NMR (100 MHz, CDCl$_3$): δ =132.2 (CH), 131.7 (CH), 119.3 (CH$_2$), 105.6 (CH), 86.5 (CH$_2$), 63.4 (CH$_2$), 55.4 (CH$_3$), 41.2 (CH$_2$), 36.2 (CH).

R$_f$ (Hex/EtOAc = 3/1): 0.12

(Z)-3-[(2R,3S,5R)-2-(tert-Butyl-diphenyl-silanyloxymethyl)-5-methoxy-tetrahydro-furan-3-yl]-acrylic acid methyl ester

18-Crown-6 (2.41 g, 9.12 mmol) and Still-Gennari phosphonate **196** (0.58 g, 1.82 mmol) were dissolved in 20 mL of THF and cooled to -78 °C. KHMDS in toluene (3.05 mL, 1.52 mmol, 0.5M) was added dropwise, followed by **207a** (0.61 g, 1.52 mmol) dissolved in 5 mL of THF. After stirring at -78 °C for 1 h, the reaction was quenched by addition of 1 mL of MeOH. After warming to r.t., sat. NaHCO$_3$ solution was added and the mixture was extracted 3 times with diethyl ether. The combined organic layers were washed with brine, dried over MgSO$_4$, filtered and concentrated under reduced pressure. Purification by column chromatography (Hex:EtOAc = 10:1) yielded **306** as a colourless oil. (0.44 g, 63%)

^{1}H NMR (250 MHz, CDCl$_3$): δ = 7.68 (m, 4 H), 7.39 (m, 6 H), 6.27 (dd, J = 11.4, 9.5 Hz, 1 H), 5.74 (dd, J = 11.4, 1.0 Hz, 1 H), 5.09 (dd, J = 5.2, 2.1 Hz, 1 H), 4.07-3.90 (m, 2 H), 3.78 (dd, J = 11.0, 4.7 Hz, 1 H), 3.72 (dd, J = 11.0, 4.3 Hz, 1 H), 3.66 (s, 3 H), 3.37 (s, 3 H), 2.48 (ddd, J = 13.4, 9.4, 5.2 Hz, 1 H), 1.70 (ddd, J = 13.4, 5.4, 2.1 Hz, 1 H), 1.05 (s, 9 H).

^{13}C NMR (63 MHz, CDCl$_3$): δ = 166.5 (C), 150.8 (CH), 135.7 (CH), 133.6 (C), 128.6 (CH), 127.6 (CH), 119.3 (CH), 105.4 (CH), 84.1 (CH), 65.5 (CH$_2$), 54.8 (CH$_3$), 51.1 (CH$_3$), 39.6 (CH$_2$), 39.3 (CH), 26.8 (CH$_3$), 19.2 (C).

R$_f$ (Hex/EtOAc = 1/1): 0.69

(*Z*)-3-((2*R*,3*S*,5*R*)-2-Hydroxymethyl-5-methoxy-tetrahydro-furan-3-yl)-acrylic acid methyl ester

Compound **306** (365 mg, 0.80 mmol) was dissolved in 5 mL of THF and a solution of TBAF in THF (1.04 mL, 1.04 mmol, 1.0 M) was added. After stirring at r.t. for 1 h, the reaction was quenched by addition of sat. NaHCO$_3$. The mixture was then extracted 4 times with EtOAc. The combined organic layers were washed with brine, dried over MgSO$_4$, filtered and concentrated under reduced pressure. Purification by column chromatography (Hex:EtOAc = 1:1 + 5 % NEt$_3$) yielded **307** as a colourless oil. (120 mg, 70%)

^{1}H NMR (250 MHz, CDCl$_3$): δ = 6.27 (dd, *J* = 11.3, 9.7 Hz, 1 H), 5.84 (dd, *J* = 11.3, 0.9 Hz, 1 H), 5.10 (dd, *J* = 2.3, 5.3 Hz, 1 H), 3.95 (m, 1 H), 3.87 (m, 1 H), 3.78 (bd, *J* = 12.6 Hz, 1 H), 3.72 (s, 3 H), 3.65 (d, *J* = 12.6 Hz, 1 H), 3.37 (s, 3 H), 2.73 (bs, OH), 2.43 (ddd, *J* = 13.5, 9.2, 5.3 Hz, 1 H), 1.75 (ddd, *J* = 13.5, 6.0, 2.3 Hz, 1 H).
^{13}C NMR (63 MHz, CDCl$_3$): δ = 166.5 (C), 150.3 (CH), 120.3 (CH), 105.4 (CH), 82.7 (CH), 62.6 (CH$_2$), 55.0 (CH$_3$), 51.5 (CH$_3$), 40.0 (CH$_2$), 38.5 (CH).

MS (EI) *m/z*: 185 ([M-MeO]$^-$, 59), 167 (4), 153 (99), 141 (13), 125 (25), 121 (100).

HRMS(EI) calcd. For C$_9$H$_{13}$O$_4$: 185.0814, Found: 185.0812.

R$_f$ (Hex/EtOAc = 1/1): 0.17

(*E*)-3-[(2*S*,3*S*,5*R*)-5-Methoxy-3-((*1Z,3E*)-4-tributylstannanyl-buta-1,3-dienyl)-tetrahydro-furan-2-yl]-acrylic acid ethyl ester

Compound **256** (200 mg, 0.42 mmol) was dissolved in 10 mL of toluene and activated MnO$_2$ (150 mg) and stabilised Wittig-reagent **138** (176 mg, 0.50 mmol) were added. The mixture was then heated to reflux for 8 h. After cooling to r.t., the solvent was removed under reduced pressure. Purification by column chromatography (Hex:EtOAc = 10:1 + 1% NEt$_3$) yielded **324** as a colourless oil. (91 mg, 40%)

^{1}H NMR (250 MHz, CDCl$_3$): δ = 6.90 (dd, *J* = 15.7, 5.2 Hz, 1 H), 6.68 (dd, *J* = 18.6, 10.4 Hz, 1 H), 6.06 (dd, *J* = 15.6, 1.6 Hz, 1 H), 5.32 (dd, *J* = 10.4, 10.0 Hz, 1 H), 5.13 (dd, *J* = 5.5, 3.0 Hz, 1 H), 4.32 (ddd, *J* = 8.6, 5.2, 1.5 Hz, 1 H), 4.18 (q, *J* = 7.1 Hz, 2 H), 3.39 (s, 3 H), 3.05 (m, 1 H), 2.49 (ddd, *J* = 13.5, 9.4, 5.5 Hz, 1 H), 1.74 (ddd, *J* = 13.5, 7.8, 3.0 Hz, 1 H), 1.60-1.20 (m, 12 H), 0.95-0.85 (m, 18 H).

^{13}C NMR (63 MHz, CDCl$_3$): δ = 166.3 (C), 145.1 (CH), 141.2 (CH), 137.0 (CH), 134.1 (CH), 128.4 (CH), 121.4 (CH), 105.4 (CH), 81.3 (CH), 60.3 (CH$_2$), 55.3 (CH$_3$), 43.5 (CH), 40.4 (CH$_2$), 29.1 (CH$_2$), 27.2 (CH$_2$), 14.2 (CH$_3$), 13.7 (CH$_3$), 9.6 (CH$_2$).

R$_f$ (Hex/EtOAc = 10/1): 0.37

(*3aS,5aR,8S,9R,9aS,9bS*)-9-Hydroxy-8-(2-hydroxy-ethyl)-3a,5a,8,9,9a,9b-hexahydro-
3*H*-naphtho[1,2-*c*]furan-1-one

Compound **225** (25 mg, 0.09 mmol) was dissolved in 3 mL of THF, then 0.2 mL of 0.5 M HCl was added and the mixture was stirred for 1 h at 40 °C. After cooling to r.t., sat. NaHCO$_3$ solution was added until no further gas evolution was visible.
Then NaBH$_4$ (14 mg, 0.38 mmol) was added and stirring was continued at r.t. for 10 min.
The reaction mixture was now diluted with brine and then extracted 5 times with DCM. The combined organic layers were dried over MgSO$_4$ and then concentrated under reduced pressure. Purification by column chromatography (EtOAc) yielded **341** as a colourless oil. (24 mg, quant.)

^{1}H NMR (600 MHz, CDCl$_3$): δ = 5.64 (m, 2 H), 5.54 (ddd, *J* = 10.2, 2.6, 2.6 Hz, 1 H), 5.39 (ddd, *J* = 9.8, 1.9, 1.9 Hz, 1 H), 4.37 (dd, *J* = 8.9, 5.8 Hz, 1 H), 4.13 (d, 8.9 Hz, 1 H), 3.92 (ddd, *J* = 10.6, 5,3, 3.8 Hz, 1 H), 3.77 (ddd, *J* = 10.5, 9.6, 3.1 Hz, 1 H), 3.52 (dd, *J* = 7.2, 3.0 Hz, 1 H), 3.52 (dd, *J* = 7.2, 3.0 Hz, 1 H), 3.33 (dd, *J* = 11.3, 8.7 Hz, 1 H), 3.05 (m, 1 H), 2.97 (m, 1 H), 2.46 (m, 1 H), 2.24 (m, 1 H), 1.78 (m, 1 H), 1.70 (m, 1 H).

^{13}C NMR (125 MHz, CDCl$_3$): δ = 179.2 (C), 131.7 (CH), 130.0 (CH), 130.0 (CH), 128.4 (CH), 123.7 (CH), 72.2 (CH$_2$), 70.2 (CH), 61.8 (CH$_2$), 44.4 (CH), 38.7 (CH), 37.5 (CH), 33.7 (CH), 33.0 (CH).

R$_f$ (Hex/EtOAc = 1/1): 0.42

(*3aS,5aR,8S,9R,9aS,9bS*)-8-[2-(*tert*-Butyl-diphenyl-silanyloxy)-ethyl]-9-hydroxy-3a,5a,8,9,9a,9b-hexahydro-3*H*-naphtho[1,2-*c*]furan-1-one

Compound **341** (24 mg, 0.09 mmol) was dissolved in 1 mL of DCM. Then, TBPDSCl (37 mg, 0.13 mmol), NEt₃ (14 mg, 20 μl, 0.14 mmol) and DMAP (0.5 mg, 3.8 μmol) were added. After 18 h, the reaction was quenched by addition of saturated NaHCO₃. After separation of the organic layer, the aqueous layer was extracted 3 times with DCM. The combined organic layers were dried over MgSO₄ and then concentrated under reduced pressure. Purification by column chromatography (Hex:EtOAc = 1:1) yielded **342** as a colourless oil. (28 mg, 60 %).

^{1}H NMR (600 MHz, CDCl₃): δ = 7.59 (m, 4 H), 7.37 (m, 2 H), 7.33 (m, 4 H), 5.58 (m, 1 H), 5.56 (ddd, *J* = 7.5, 4.9, 2.4 Hz, 1 H), 5.49 (ddd, *J* = 10.1, 2.5, 2.5 Hz, 1 H), 5.27 (ddd, *J* = 9.7, 1.9, 1.7 Hz, 1 H), 4.31 (dd, *J* = 8.7, 6.0 Hz, 1 H), 4.06 (d, *J* = 8.7 Hz, 1 H), 3.73 (m, 2 H), 3.52 (dd, *J* = 6.9, 2.7 Hz, 1 H), 3.36 (m, 1 H), 2.94 (m, 1 H), 2.91 (m, 1 H), 2.43 (ddd, *J* = 10.9, 5.8, 2.9 Hz, 1 H), 2.25 (m, 1 H), 1.65 (m, 1 H), 1.60 (m, 1 H).

^{13}C NMR (125 MHz, CDCl₃): δ = 178.9 (C), 135.6 (CH), 135.5 (CH), 132.8 (C), 132.7 (C), 131.9 (CH), 130.0 (CH), 129.9 (CH), 128.5 (CH), 127.9 (CH), 127.8 (CH), 123.8 (CH), 72.1 (CH₂), 69.8 (CH), 63.3 (CH₂),, 44.2 (CH), 38.5 (CH), 37.4 (CH), 37.2 (CH₂), 33.7 (CH), 33.0 (CH), 26.8 (CH₃), 19.0 (C).

IR [cm^{-1}]: 3413, 3071, 3019, 2930, 2857, 1772, 1589, 1472, 1428, 1391, 1364, 1209, 1163, 1112, 1081, 1042, 1007.

MS (EI) *m/z*: 431 (M-Bu⁺, 2), 395 (7), 353 (20), 277 (39), 230 (37), 199 (45), 173 (58).

HRMS(EI) calcd. For $C_{26}H_{27}O_4Si$: 431.1679, Found: 431.1675.

R_f (Hex/EtOAc = 1/1): 0.73

(*3aS,5aR,8S,9R,9aS,9bS*)-8-[2-(*tert*-Butyl-diphenyl-silanyloxy)-ethyl]-9-triethylsilanyloxy-3a,5a,8,9,9a,9b-hexahydro-3*H*-naphtho[1,2-*c*]furan-1-one

Compound **342** (23 mg, 0.04 mmol) was dissolved in 1 mL DCM. At 0 °C, NEt_3 (13 mg, 17 µl, 0.13 mmol) and TESOTf (25 mg, 22 µl, 0.10 mmol) was added. After warming to r.t., the reaction was stirred for 4 hours, before it was quenched by addition of saturated NH_4Cl-solution. After separation of the organic layer, the aqueous layer was extracted 4 times with DCM. The combined organic layers were dried over $MgSO_4$ and then concentrated under reduced pressure. Purification by column chromatography (Hex:EtOAc = 7:1) yielded **343** as a colourless oil. (26 mg, 90 %)

^{1}H NMR (600 MHz, $CDCl_3$) δ = 7.66 (m, 4 H), 7.42 (m, 2 H), 7.38 (m, 4 H), 5.63 (bd, *J* = 9.8 Hz, 1 H), 5.56-5.49 (m, 3 H), 4.33 (dd, *J* = 8.9, 5.8 Hz, 1 H), 4.33 (d, *J* = 8.9 Hz, 1 H), 3.75 (ddd, *J* = 10.4, 7.1, 4.8 Hz, 1 H), 3.72 (ddd, *J* = 10.4, 7.9, 6.0 Hz, 1 H), 3.42 (dd, *J* = 10.6, 8.3 Hz, 1 H), 3.23 (dd, *J* = 6.9, 3.1 Hz, 1 H), 2.96 (m, 1 H), 2.48 (ddd, *J* = 10.6, 5.7, 3.3 Hz, 1 H), 2.33 (m, 1 H), 2.14 (m, 1 H), 1.33 (m, 1 H), 1.04 (s, 9 H), 0.98 (t, *J* = 8.0 Hz, 9 H), 0.66 (q, *J* = 8.0 Hz, 6 H).

^{13}C NMR (125 MHz, $CDCl_3$) δ = 178.6 (C), 135.6 (CH), 135.6 (CH), 133.9 (C), 133.8 (C), 132.3 (CH), 129.6 (CH), 129.5 (CH), 128.4 (CH), 127.7 (CH), 127.6 (CH), 127.3 (CH), 123.1 (CH), 72.1 (CH), 71.9 (CH_2), 61.4 (CH_2), 41.9 (CH), 38.9 (CH), 38.4 (CH), 34.8 (CH_2), 33.8 (CH), 33.3 (CH), 26.8 (CH_3), 19.2 (C), 7.2 (CH_3), 6.1 (CH_2).

IR [cm⁻¹]: 3020, 2957, 2930, 2857, 1589, 1472, 1728, 1362, 1209, 1161, 1113, 1083, 1042.

MS (EI) m/z: 574 (M-Et⁺, 41), 545 (100), 515 (12), 467 (9), 413 (6), 313 (14), 215 (10).

HRMS(EI) calcd. For $C_{34}H_{46}O_4Si_2$: 574.2935, Found: 574.2927.

R_f (Hex/EtOAc = 3/1): 0.52

(*1S,2S,4aR,7S,8R,8aS*)-7-[2-(*tert*-Butyl-diphenyl-silanyloxy)-ethyl]-2-hydroxymethyl-8-triethylsilanyloxy-1,2,4a,7,8,8a-hexahydro-naphthalene-1-carboxylic acid methoxy-methyl-amide

Compound **343** (23 mg, 0.04 mmol) was dissolved in 2.5 mL THF and MeNHOMe·HCl (19 mg, 0.19 mmol) was added. At 0 °C, isopropymagnesium chloride in THF (0.19 mL, 0.38 mmol, 2 M) was added. After warming to r.t., the reaction was stirred for 1 h. Then a saturated NH₄Cl-solution was added, and the reaction mixture was extracted 4 times with DCM. The combined organic layers were dried over MgSO₄ and then concentrated under reduced pressure. Purification by column chromatography (Hex:EtOAc = 1:1) yielded **347** as a colourless oil. (21.5 mg, 85 %)

¹H NMR (600 MHz, CDCl₃) δ = 7.65 (m, 4 H), 7.42 (m, 2 H), 7.37 (m, 4 H), 5.81 (ddd, *J* = 9.9, 4.1, 1.7 Hz, 1 H), 5.59 (ddd, *J* = 10.0, 4.2, 2.3 Hz, 1 H), 5.50 (m, 2 H), 3.85 (m, 1 H), 3.75-3.65 (m, 6 H), 3.56 (m, 1 H), 3.41 (m, 1 H), 3.16 (m, 3 H), 2.59 (m, 1 H),

2.40 (m, 1 H), 2.37 (m, 1 H), 2.18 (m, 1 H), 1.74 (m, 1 H), 1.04 (s, 9 H), 0.95 (t, J = 7.9 Hz, 9 H), 0.65 (q, J = 7.9 Hz, 6 H).

^{13}C NMR (62 MHz, CDCl$_3$) δ = 175.5 (C), 135.5 (CH), 133.8 (C), 131.0 (CH), 129.6 (CH), 127.6 (CH), 127.0 (CH), 126.3 (CH), 125.2 (CH), 71.2 (CH$_3$), 63.3 (CH$_2$), 62.0 (CH$_2$), 39.8 (CH), 39.3 (CH), 37.9 (CH), 36.8 (CH), 35.8 (CH$_2$), 32.7 (CH$_3$), 32.5 (CH), 26.8 (CH$_3$), 19.1 (C), 7.0 (CH$_3$), 5.2 (CH$_2$).

R$_f$ (Hex/EtOAc = 1/1): 0.27

(*1aR,2S,3R,3aS,4R,5S,7aR,7bS*)-5-[2-(*tert*-Butyl-diphenyl-silanyloxy)-ethyl]-2-hydroxymethyl-4-triethylsilanyloxy-1a,2,3,3a,4,5,7a,7b-octahydro-1-oxa-cyclopropa[*a*]naphthalene-3-carboxylic acid methoxy-methyl-amide

Compound **347** (16.5 mg, 0.02 mmol) was dissolved in 3 mL DCM. Then 1 mL of phosphate buffer (pH 7) was added, followed by *m*CPBA (7.4 mg, 0.03 mmol, 77%). After stirring for 40 min, the reaction was quenched by addition of 1 mL of 1 M Na$_2$S$_2$O$_3$, and 5 mL of saturated NH$_4$Cl. The mixture was extracted 4 times with DCM. The combined organic layers were dried over MgSO$_4$ and then concentrated under reduced pressure. Purification by column chromatography (Hex:EtOAc = 1:1) yielded **348** as a colourless oil. (11 mg, 65 %)

^{1}H NMR (250 MHz, CDCl$_3$) δ = 7.64 (m, 4 H), 7.39 (m, 6 H), 5.70 (d, J = 10.5 Hz, 1 H), 5.53 (d, J = 10.5 Hz, 1 H), 3.99 (m, 1 H), 3.82 (m, 1 H), 3.75-3.65 (m, 5 H), 3.34 (m, 1 H), 3.28 (m, 1 H), 3.10 (s, 3 H), 2.96 (m, 1 H), 2.60 (m, 1 H), 2.24 (m, 1 H), 2.18 (m, 1 H), 1.03 (s, 9 H), 0.93 (t, J = 7.5 Hz, 9 H), 0.61 (q, J = 7.5 Hz, 6 H).

R_f (Hex/EtOAc = 1/1): 0.10

Epoxide Lacton

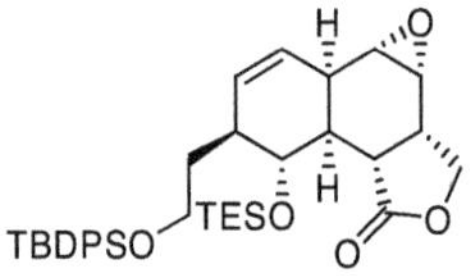

350

^{1}H NMR (400 MHz, CDCl$_3$) δ = 7.66 (m, 4 H), 7.40 (m, 6 H), 5.64 (ddd, J = 10.2, 2.5, 2.5 Hz, 1 H), 5.58 (ddd, J = 10.2, 3.4, 2.6 Hz, 1 H), 4.42 (m, 2 H), 3.75 (m, 1 H), 3.65 (dd, J = 8.1, 5.8 Hz, 1 H), 3.23 (dd, J = 3.5, 3.5 Hz, 1 H), 3.09 (dd, J = 3.6, 0.1 Hz, 1 H), 2.94 (m, 1 H), 2.84 (m, 1 H), 2.76 (m, 1 H), 2.36 (m, 1 H), 2.18 (m, 1 H), 2.10 (m, 1 H), 1.26 (m, 1 H), 1.04 (s, 9 H), 0.96 (t, J = 7.8 Hz, 9 H), 0.62 (q, J = 7.8 Hz, 6 H).

^{13}C NMR (100 MHz, CDCl$_3$) δ = 177.3 (C), 135.6 (CH), 135.5 (CH), 133.8 (C), 130.6 (CH), 129.6 (CH), 129.5 (CH), 127.7 (CH), 127.6 (CH), 123.7 (CH), 71.7 (CH), 69.0 (CH$_2$), 61.5 (CH$_2$), 55.3 (CH), 51.6 (CH), 41.0 (CH), 35.5 (CH$_2$), 34.9 (CH), 34.7 (CH), 32.6 (CH), 31.0 (CH), 26.8 (CH$_3$), 19.2 (C), 7.1 (CH$_3$), 5.6 (CH$_2$).

R_f (Hex/EtOAc = 3/1): 0.42

(*1aR,2S,3R,3aS,4R,5S,7aR,7bS*)-5-[2-(*tert*-Butyl-diphenyl-silanyloxy)-ethyl]-2-methoxymethyl-4-triethylsilanyloxy-1a,2,3,3a,4,5,7a,7b-octahydro-1-oxa-cyclopropa[*a*]naphthalene-3-carboxylic acid methoxy-methyl-amide

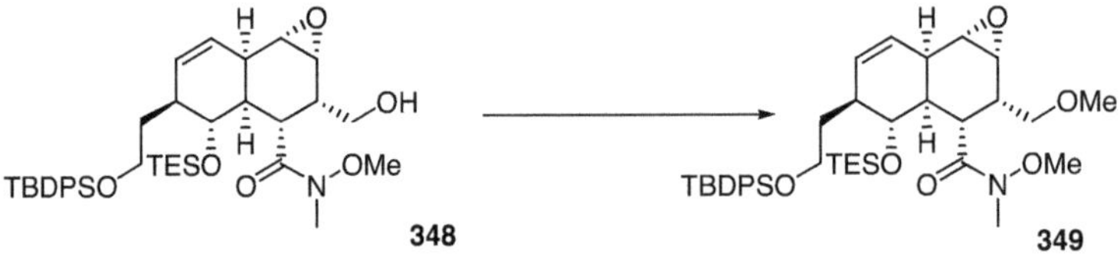

348 → **349**

Compound **348** (1.5 mg, 2.2 µmol) was dissolved in 1 mL THF. At 0 °C, sodium hydride (0.25 mg, 6.6 µmol) was added, followed by MeI (3.2mg, 1.5µl, 22 µmol) after 10 minutes. After warming to r.t., stirring continued for 2 h.

The reaction was quenched by addition of saturated NH_4Cl solution and then extracted 4 times with DCM. The combined organic layers were dried over $MgSO_4$ and then concentrated under reduced pressure. Purification by column chromatography (Hex:EtOAc = 3:1) yielded **349** as a colourless oil. (1 mg, 65 %)

^{1}H NMR (600 MHz, $CDCl_3$) δ = 7.65 (m, 4 H), 7.42 (m, 2 H), 7.38 (m, 4 H), 5.69 (m, 1 H), 5.52 (d, *J* = 10.3 Hz, 1 H), 4.03 (m, 1 H), 3.67 (dd, *J* = 6.9, 6.7 Hz, 1 H), 3.65 (s, 3 H), 3.59 (dd, *J* = 9.5, 6.7 Hz, 1 H), 3.33 (m, 1 H), 3.31 (dd, *J* = 5.4, 3.8 Hz, 1 H), 3.29 (s, 3 H), 3.27 (m, 1 H), 3.07 (s, 3 H), 2.89 (m, 1 H), 2.69 (m, 1 H), 2.16 (m, 2 H), 2.01 (m, 1 H), 1.65 (m, 1 H), 1.57 (m, 1 H), 1.16 (m, 1 H), 1.03 (s, 9 H), 0.93 (t, *J* = 7.8 Hz, 9 H), 0.61 (q, *J* = 7.8 Hz, 6 H).

11.4 Side Chain Synthesis

[(*S,S*)-5-(Hydroxymethyl)-2,2-dimethyl-1,3-dioxolan-4-yl]methanol[159]

In a 1 L single-neck flask, (*R,R*)-dimethyltartrate **373** (54.45 g, 264 mmol) was dissolved in dry DCM (500 mL) and 2,2-dimethoxypropane (53 mL) was added. Then *p*-toluenesulfonic acid was added until pH~3 was reached. A soxhlet apparatus filled with 3 Å molecular sieves was installed and the mixture was heated to reflux for 4 hours. The reaction was cooled and quenched with sat. $NaHCO_3$ solution; DCM was removed under vacuum and the mixture was extracted with diethyl ether. The organic layer was dried over $MgSO_4$, filtered and concentrated. The residue was subjected to distillation under vacuum (bp 100/102 °C, 0.5 mbar) to afford the pure acetonide diester. (48.96 g, 85%)

^{1}H NMR (250 MHz, $CDCl_3$): δ = 4.74 (s, 2 H), 3.77 (s, 6 H), 1.43 (s, 6 H). *cf.* Ref. 159

A solution of $LiAlH_4$ (13 g, 342 mmol) in anhydrous Et_2O (400 mL) was prepared and cooled to -10 °C. A solution of the acetonide diester (47 g, 215 mmol) in Et_2O was added dropwise and then the mixture was stirred at r.t. for 4 hours. A solution of $MgSO_4$ was then added and the mixture was stirred for another hour; after filtration through a pad of celite, diglyme (100 mL) was added and the slurry was heated to reflux for one hour. The mixture was then filtered, and the filtrate dried over $MgSO_4$, filtered and concentrated. The residue was subjected to distillation under vacuum (bp 100/102 °C, 0.6 mbar) to afford pure **375**. (24.4 g, 70%).

---

[159] Khanapure, S.P.; Najafi, N.; Manna, S.; Yang, J.; Rokach J. *J.Org.Chem.* **1995**, *23*, 7548-7551.

^{1}H NMR (250 MHz, CDCl$_3$): δ 4.02 (m, 2 H), 3.75 (m, 4 H), 2.28 (br s, 2 H), 1.42 (s, 6 H). *cf*. Ref. 159

[(4*S*,5*S*)-5-(Methoxymethyl)-2,2-dimethyl-1,3-dioxolan-4-yl]methanol

NaH (4 g of 60% dispersion in mineral oil, 100 mmol) was dissolved in 100 mL of DMF and cooled to -20 °C when **374** (17.3 g, 107 mmol) dissolved in 20 mL of DMF was added. The mixture was then warmed to 0 °C and stirred for 30 minutes. MeI (5.6 mL, 90 mmol) was added dropwise and the mixture warmed to r.t. and stirred for one hour. The reaction was then quenched with ice and extracted three times with EtOAc. The combined organic layers were washed with NaHCO$_3$ and brine, dried over MgSO$_4$, filtered and concentrated under reduced pressure. Purification by column chromatography (Hex:EtOAc = 1:1) yielded **375** as a colourless oil. (12.23 g, 65%).

^{1}H NMR (400 MHz, CDCl$_3$): δ 4.01 (ddd, J = 8.3, 5.3, 5.3 Hz, 1 H), 3.88 (ddd, J = 8.3, 5.0, 5.0 Hz, 1 H), 3.62-3.65 (m, 2 H), 3.59 (dd, J = 10.0, 5.0 Hz, 1 H), 3.51 (dd, J = 10.0, 5.0 Hz, 1 H), 3.43 (s, 3 H), 2.26 (brs, 1 H), 1.44 (s, 6 H).

^{13}C NMR (100 MHz, CDCl$_3$): δ 107.1 (C), 77.0 (CH), 74.1 (CH), 70.7 (CH$_2$), 59.9 (CH$_3$), 57.0 (CH$_2$), 24.43 (CH$_3$).

IR [cm^{-1}]: 3455 (br), 2987, 2933, 1747, 1653, 1559, 1456, 1372, 1251, 1168, 1086.

MS(EI) *m/z* 161 (M-15, 27), 145 (9), 131 (10), 87 (11).

HRMS(EI) calcd. For C$_7$H$_{13}$O$_4$: 161.0814, Found: 161.0812.

$[\alpha]_D^{20}$ +5.2 (c 1.6, CH$_2$Cl$_2$).

R$_f$ (Hex/EtOAc = 1/1): 0.48

For data of the enantiomer, see Ref. 160

[(4*S*,5*S*)-5-(Methoxymethyl)-2,2-dimethyl-1,3-dioxolan-4-yl]methyl-4-methyl-benzenesulfonate

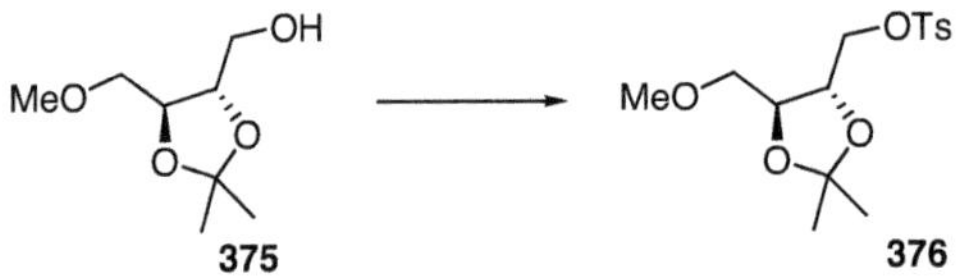

To a stirred solution of **375** (7g, 40 mmol) in pyridine (20 mL) at 0 °C, *p*-toluenesulfonyl chloride (7.6 g, 40 mmol) was added slowly, followed by a catalytic amount of DMAP was then added. The reaction was warmed to r.t. and stirred for 1 hour.

The mixture was concentrated under reduced pressure, dissolved with DCM and was then washed with 1M HCl, sat. NaHCO$_3$ and brine, dried over MgSO$_4$, filtered and concentrated. Purification by column chromatography (Hex:EtOAc = 3:1) yielded **376** as a colourless oil. (11.55 g, 88 %)

^{1}H NMR (400 MHz, CDCl$_3$): δ 7.81 (d, *J* = 8.4 Hz, 2 H), 7.36 (d, *J* = 8.4 Hz, 2 H), 4.16 (m, 1 H), 4.08 (m, 1 H), 3.99 (m, 2 H), 3.50 (m, 2 H), 3.37 (s, 3 H), 2.46 (s, 3 H), 1.38 (s, 3 H), 1.34 (s, 3 H).

---

[160] Kessinger, R.; Thilgen, C.; Mordasini, T.; Diederich, F. *Helv. Chim. Acta* **2000**, *83*, 3069-3096

^{13}C NMR (100 MHz, CDCl$_3$): δ 145.4 (C), 133.2 (C), 130.2 (CH), 128.4 (CH), 110.6 (C), 76.3 (CH), 76.0 (CH), 73.1 (CH$_2$), 69.6 (CH$_2$), 59.9 (CH$_3$), 27.3 (CH$_3$), 27.1 (CH$_3$), 22.0 (CH$_3$).

IR [cm^{-1}]: 2987, 2931, 1925, 1746, 1652, 1598, 1495, 1455, 1361, 1308, 1292, 1178, 1096.

MS(EI) m/z 315 (M-15, 16), 285 (3), 227 (8), 190 (5), 155 (37), 146 (11), 99 (20), 91 (100).

HRMS(EI) calcd. For C$_{14}$H$_{19}$O$_6$S: 315.0902, Found: 315.0900.

[α]$_D^{20}$ -12.4 (c 6.0, CHCl$_3$).

R$_f$ (Hex/EtOAc = 3/1): 0.40

[(4*S*,5*R*)-5-(Iodomethyl)-2,2-dimethyl-1,3-dioxolan-4-yl]methyl methyl ether

To a stirred solution of tosylate **376** (9 g, 27.3 mmol) in CH$_3$CN (80 mL), NaI (6,14g, 40.9 mmol) was added and the mixture was heated at reflux for 5 hours. The solid was then filtered off and the solvent was removed under vacuum. Diethyl ether was added and the organic layer was washed with Na$_2$S$_2$O$_3$, water and brine. The organic layer was dried over MgSO$_4$, filtered and concentrated. Purification by column chromatography (Hex:EtOAc = 5:1) yielded **377** as a colourless oil. (7.25 g, 93%).

^{1}H NMR (400 MHz, CDCl$_3$): δ = 3.94 (m, 1 H), 3.83 (m, 1 H), 3.59 (d, *J* = 5.0 Hz, 2 H), 3.42 (s, 3 H), 3.32 (ddd, *J* = 15.8, 10.5, 5.0 Hz, 2 H), 1.47 (s, 3 H), 1.43 (s, 3 H).

^{13}C NMR (100 MHz, CDCl$_3$): δ 110.2 (C), 80.4 (CH$_2$), 77.9 (CH), 73.7 (CH$_2$), 59.9 (CH$_3$), 27.8 (CH$_3$), 27.7 (CH$_3$), 6.6 (CH$_2$).

IR [cm^{-1}]: 2986, 2932, 1748, 1653, 1559, 1456, 1414, 1380, 1371, 1321, 1238, 1161, 1104.

MS(EI) m/z 271 (M-15, 100), 241 (21), 211 (14), 185 (18), 183 (19), 155 (11), 127 (22).

HRMS(EI) calcd. For C$_7$H$_{12}$O$_3$I: 270.9831, Found: 270.9824 .

[α]$_D^{20}$ -11.8 (c 3.3, CHCl$_3$).

R$_f$ (Hex/EtOAc = 3/1): 0.74

(*R*)-Glycidol-methylether

HO—△O ⟶ MeO—△O
380   381

(*S*)-Glycidol **380** (1.0g, 13.50 mmol) was dissolved in 10 mL of dry DCM. Then Ag$_2$O (3.13 g, 13.50 mmol) and MeI (8.40 mL, 135.0 mmol) was added and the resulting suspension was heated to reflux for 24 h.

After cooling to r.t., the mixture was filtered through a pad of celite, which was then washed twice with 10 mL of dichloromethane. From this solution, removal of MeI and DCM by fractional distillation yielded pure **381**. (0.93 g, 78 %)

^{1}H NMR (CDCl$_3$, 250 MHz): δ = 3.69 (dd, *J* = 11.4, 2.9 Hz, 1 H), 3.41 (s, 3 H), 3.34 (dd, *J* = 11.4, 5.8 Hz, 1 H), 3.16 (m, 1 H), 2.81 (dd, *J* = 9.1, 4.2 Hz), 2.61 (dd, 5.0, 2.7 Hz).

^{13}C NMR (CDCl$_3$, 75 MHz): δ = 73.1 (CH$_2$), 59.1 (CH$_3$), 50.7 (CH), 44.1 (CH$_2$).

[α]$_D^{20}$ -15.0 (c 5.0, toluene).

(R)-1-Methoxy-3-buten-2-ol (Method A):

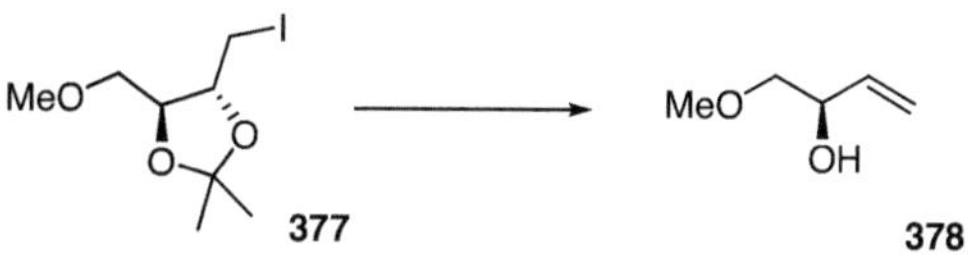

To a stirred solution of the dioxolan-methyliodide **377** (720 mg, 2.5 mmol) in EtOH (16 mL), activated Zn (1.16 g, 17.7 mmol) was added and the mixture was stirred under argon at reflux for 5 hours. The solid was then filtered through a pad of celite and the solvent was removed under vacuum. The residue was dissolved in EtOAc, washed with water and brine, dried over MgSO$_4$, filtered and concentrated under reduced pressure. Purification by column chromatography (Hex:EtOAc = 3:1) yielded **378** as a colourless oil. (180 mg, 70%).

^{1}H NMR (400 MHz, CDCl$_3$): δ = 5.84 (ddd, J = 17.3, 10.5, 5.6 Hz, 1 H), 5.36 (ddd, J =17.3, 1.6, 1.6 Hz), 5.20 (ddd, 10.5, 1.6, 1.5 Hz, 1 H), 4.31 (m, 1 H), 3.45 (dd, J = 9.6, 3.5 Hz, 1 H), 3.30 (dd, 9.6, 7.8 Hz, 1 H).

^{13}C NMR (100 MHz, CDCl$_3$): δ = 137.1 (CH), 116.8 (CH$_2$), 76.8 (CH$_2$), 71.7 (CH), 59.5 (CH$_3$).

IR [cm^{-1}]: 3423 (br), 2925, 2857, 1746, 1653, 1559, 1447, 1374, 1247, 1106, 1037.

[α]$_D^{20}$ +4.7 (c 1.0, CHCl$_3$).

R$_f$ (Hex/EtOAc = 1/1): 0.26

(R)-1-Methoxy-3-buten-2-ol (Method B)

MeO—⟨O⟩ → MeO—CH(OH)—CH=CH₂
   381   **378**

Trimethylsulfonium iodide (4.95 g, 24.2 mmol) was suspended in 30 mL of toluene and heated to reflux for 30 min on a Dean-Stark trap. Toluene was then removed under reduced pressure and the residue dried under high vacuum.

After flushing the flask with argon, 50 mL of anhydrous THF were added, the suspension cooled to -15°C with an ice/salt bath and *n*-BuLi in hexanes (9.7 mL, 24.2 mmol, 2.5 M) was added over 30 min. After stirring at -15°C for another 30 min, (*R*)-glycidol-methylether **381** (712 mg, 8.08 mmol) dissolved in 10 mL of THF was added within 15 minutes. The mixture was stirred at -10 °C for another hour and then warmed to r.t. overnight.

The reaction was quenched by the addition of 30 mL water and then extracted 4 times with diethyl ether. The combined organic layers were washed with brine, dried over MgSO₄, filtered and carefully concentrated under reduced pressure. Purification by column chromatography (Hex:EtOAc = 3:1) yielded **378** as a colourless oil. (627 mg, 76%)

Spectroscopic analysis was identical to that of **378** obtained by Method A

$[\alpha]_D^{20}$ +4.5 (c 1.0, CHCl₃).

Triisopropyl[(*R*)-1-(methoxymethyl)-2-propenyl]oxysilane

MeO—CH(OH)—CH=CH₂ → MeO—CH(OTIPS)—CH=CH₂
     378   **379**

Compound **378** (255 mg, 2.5 mmol) was dissolved in 5 mL of DMF and imidazole (510 mg, 7.5 mmol) and triisopropylsilyl chloride (722 mg, 0.80 mL, 3.7 mmol) were added. The mixture was stirred for 14 hours. Diethyl ether was added then to the reaction and the mixture was washed with 0.5 M KHSO$_4$ (1x2 mL), water and brine, dried over MgSO$_4$, filtered and concentrated. Purification by column chromatography (Hex:EtOAc 50:1 to 20:1) yielded **379** as a colourless oil. (594 mg, 92%).

^{1}H NMR (400 MHz, CDCl$_3$): δ = 5.90 (ddd, J = 17.2, 10.4, 5.8 Hz, 1 H), 5.33 (d, J = 17.2 Hz, 1 H), 5.17 (d, J = 10.4 Hz, 1 H), 4.38 (m, 1 H), 3.57-3.30 (m, 2 H), 3.38 (s, 3 H), 1.08 (m, 21 H).

^{13}C NMR (100 MHz, CDCl$_3$): δ 139.7 (CH), 115.5 (CH$_2$), 78.1 (CH$_2$), 73.2 (CH), 59.5 (CH$_3$), 18.1 (CH$_3$), 12.8 (CH).

IR [cm^{-1}]: 2944, 2867, 1700, 1684, 1653, 1647, 1559, 1464, 1420, 1402, 1383, 1366, 1340, 1247, 1197, 1126, 1104, 1035, 1014.

MS(EI) m/z 215 (M-43, 47), 187 (22), 173 (2), 157 (10), 145 (100), 131 (12), 117 (67).

HRMS(EI) calcd. For C$_{11}$H$_{22}$O$_2$Si: 215.1462, Found: 215.1462.

[α]$_D^{20}$ +14.0 (c 0.2, CHCl$_3$).

R$_f$ (Hex/EtOAc = 10/1): 0.42

(2S)-3-Methoxy-2-[(1,1,1-triisopropylsilyl)oxy]propanal

Compound **379** (144 mg, 0.56 mmol) was dissolved in DCM (15 mL) and cooled to −78 °C. O_3 enriched air was bubbled through the reaction mixture until a faint blue colour persisted. PPh$_3$ (160 mg, 0.61 mmol) was then added and the mixture was stored at -18 °C overnight. After addition of 400 mg of SiO_2, the solvent was carefully removed under reduced pressure. Loading the silica adsorbed material on a column (preconditioned and loaded with 6g of SiO_2) and fast chromatography (pentane : Et_2O = 20:1 to 7:1) yielded **369** as a colourless oil. (131 mg, 90 %).

^{1}H NMR (250 MHz, CDCl$_3$): δ = 9.72 (s, 1 H), 4.23 (t, J = 4.8 Hz, 1 H), 3.65 (dd, J = 4.8, 2.5 Hz, 2 H), 3.38 (s, 3 H), 1.09 (m, 21 H).

^{13}C NMR (CDCl$_3$): δ = 203.6 (C), 77.7 (CH), 74.5 (CH$_2$), 59.9 (CH$_3$), 18.2 (CH$_3$), 12.7 (CH).

IR [cm^{-1}]: 3056, 2925, 2866, 1740 1701, 1653, 1617, 1465, 1437, 1381, 1306, 1247, 1120, 1028.

R$_f$ (Hex/EtOAc = 10/1): 0.34

(R)-4-trimethylsilyl-3-butyn-2-yl acetate and (S)-4-trimethylsilyl-3-butyn-2-ol[144]

Amano lipase AK (1g) and 4 Å molecular sieves (530 mg) were suspended in pentane (20 mL) and stirred for 5 min. To this suspension were added 4-trimethylsilyl-3-butyn-2-ol **383** (4.9 g, 35 mmol) and freshly distilled vinyl acetate (26 mL) and the mixture was stirred for 24 hours. The mixture was then filtered through celite, dried (MgSO$_4$) and concentrated under reduced pressure. Purification by column chromatography (pentane:diethyl ether = 8:1) yielded **384** and (*R*)-4-trimethylsilyl-3-butyn-2-yl acetate **385** as colourless oils. **384:** (2.20 g, 45 %)

Data for **384**. ^{1}H NMR (250 MHz, CDCl$_3$): δ 4.54 (q, *J* = 6.5, 1 H), 1.84 (bs, 1 H), 1.46 (d, *J* = 6.6 Hz, 3 H), 0.19 (s, 9 H). *cf*. Ref. 144

(*S*)-4-Trimethylsilyl-3-butyn-2-yl mesylate[140c]

Propargylic alcohol **384** (2.3g, 16.2 mmol) was dissolved in DCM (90 mL) and cooled to -78 °C. Et$_3$N (3.4 mL, 24.3 mmol) and MsCl (1.24 mL, 16.2 mmol) was added and the solution was stirred for one hour at -78 °C.
The reaction was quenched at this temperature by addition of sat. NaHCO$_3$. The mixture was allowed to warm to r.t. After separation of the organic layer, it was dried over MgSO$_4$, filtered and concentrated under reduced pressure to yield **386** as a colourless oil which was employed without further purification for the next step.

^{1}H NMR (CDCl$_3$): δ 5.29 (q, *J* = 6.8 Hz, 1 H), 3.13 (s, 3 H), 1.65 (d, *J* = 6.8 Hz, 3 H), 0.21 (s, 9 H). *cf*. Ref.140c

(P)-2-trimethylsilylpenta-2,3-diene[144]

A single-neck flask was charged with CuBr (1.322 g, 9.21 mmol) and LiBr (9.21 g, 9.21 mmol) and heated to 120 °C in an oil bath under high vacuum overnight. The flask was then cooled and freshly distilled THF (14 mL) was added; the resulting solution was cooled to 0 °C and MeMgCl (3.1 mL of a 3M solution in THF) was added. The resulting pale green paste was stirred for 20 min and then cooled to -78 °C. A solution of **386** (1.13 g, 5.12 mmol) in 4 mL of THF was slowly added and the resulting mixture was stirred at -78 °C for 40 min, warmed to r.t. and stirred for a further 2 hour. The mixture was then poured into a mixture of pentane (40 mL), water (20 mL) and saturated NH_4Cl solution (40 mL) and stirred for 30 min. The organic layer was then separated, washed with water, brine, dried ($MgSO_4$), filtered and carefully concentrated under reduced pressure. Purification by flash chromatography (pentane) yielded **8** as a colourless oil. (643 mg, 90%)

^{1}H NMR (CDCl$_3$): δ 4.68 (qq, J = 6.9, 2.8 Hz, 1 H), 1.65 (d, J = 2.8 Hz, 3 H), 1.60 (d, J = 6.9 Hz, 3 H), 0.09 (s, 9 H).

$[\alpha]_D^{20}$ +77 (*c* 1.2, CHCl$_3$). *cf.* Ref. 144

(2S,3S,4R)-1-methoxy-4-methyl-2-[(1,1,1-triisopropylsilyl)oxy]-5-heptyn-3-ol

A mixture of aldehyde **369** (150 mg, 0.5 mmol) and allene **370** (105 mg, 0.75 mmol) in anhydrous DCM (6 mL) was cooled to -78 °C. TiCl$_4$ (0.5 mL of a 1 M solution in DCM) was then added. After 2 hours the solution was warmed to -20°C and quenched with a saturated NH$_4$Cl solution. After warming to r.t. the solvent was removed under vacuum. The residue was redissolved in diethyl ether and then washed with water and brine, dried over MgSO$_4$, filtered and concentrated. Purification by column chromatography (Hex : EtOAc = 10:1) yielded **387** as a colourless oil (mixture of diasteroisomers syn : anti = 20:1). (146 mg, 77%).

^{1}H NMR (CDCl$_3$): δ = 4.44 (ddd, J = 6.8, 5.6, 1.0 Hz, 1 H), 3.48 (dd, J = 9.3, 6.8 Hz, 1 H), 3.41 (dd, J = 9.3, 5.6 Hz, 1 H), 3.34 (s, 3 H), 3.31 (m, 1 H), 2.49 (m, 1 H), 2.34 (bd, 5.3 Hz, 1 H), 1.76 (d, J = 2.5 Hz, 3 H), 1.26 (d, J = 6.8 Hz, 3 H), 1.15 – 1.05 (m, 21 H).

^{13}C NMR (CDCl$_3$): δ = 81.4 (C), 78.5 (C), 75.6 (CH), 75.1 (CH$_2$), 71.4 (CH), 59.3 (CH$_3$), 30.8 (CH), 18.5 (CH$_3$), 18.2 (CH), 13.4 (CH$_3$), 3.9 (CH$_3$).

IR [cm^{-1}]: 3403 (br), 2926, 2867, 1749, 1559, 1462, 1384, 1289, 1199, 1123, 1061, 1010.

MS(EI) m/z 285 (M-43, 11), 267 (15), 261 (18), 253 (32), 215 (8), 187 (43), 145 (57).

HRMS(EI) calcd. For C$_{15}$H$_{29}$O$_3$Si: 285.1886, Found: 285.1879 .

[α]$_D^{20}$ +19.8 (c 2.0, CHCl$_3$).

R$_f$ (Hex/EtOAc = 10/1): 0.39

(4R,5S,6S)-5-Hydroxy-7-methoxy-4-methyl-6-triisopropylsilanyloxy-heptan-2-one

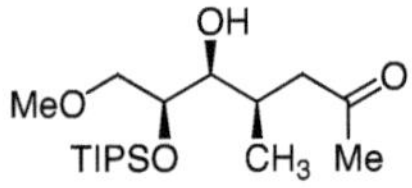

388

^{1}H NMR (400 MHz, CDCl$_3$): δ = 3.91 (ddd, J = 5.6, 5.2, 4.4 Hz, 1 H), 3.46 (dd, J = 10.2, 4.4 Hz, 1 H), 3.38 (dd, J = 10.2, 5.2 Hz, 1 H), 3.32 (s, 3 H), 2.65 (dd, J = 17.3, 5.3 Hz, 1 H), 2.59 (bd, J = 4.3 Hz, OH), 2.40 (dd, J = 17.3, 8.5 Hz, 1 H), 2.23 (m, 1 H), 2.13 (s, 3 H), 1.10 (m, 21 H), 0.93 (d, J = 6.6 Hz, 3 H).

^{13}C NMR (100 MHz, CDCl$_3$): δ = 208.4 (C), 74.7 (CH$_2$), 74.5 (CH), 72.9 (CH), 58.9 (CH$_3$), 48.3 (CH$_2$), 30.7 (CH$_3$), 30.6 (CH), 18.2 (CH$_3$), 18.1 (CH$_3$), 12.8 (CH).

IR [cm^{-1}]: 3543 (br), 2943, 2892, 2867, 2812, 1717, 1464, 1404, 1383, 1362, 1247, 1197, 1123, 1066.

MS (EI) m/z 303 (M$^+$-43, 31), 285 (94), 271 (19), 253 (42), 211 (17), 187 (21), 157 (18), 145 (62).

HRMS(EI) calcd. For C$_{18}$H$_{38}$O$_4$Si: 303.1982, Found: 303.1992.

R$_f$ (Hex/EtOAc = 10/1): 0.08

((1S,2S,3R)-2-Benzyloxy-1-methoxymethyl-3-methyl-hex-4-ynyloxy)-triisopropyl-silane

387 → **366**

NaH (14.4 mg of 60% dispersion in mineral oil) was dissolved in DMF (3 mL) and cooled to -20 °C. Compound **387** (80 mg, 0.24 mmol) dissolved in DMF (1 mL) was then added, warmed to 0 °C and stirred for 30 minutes. Then BnBr (0.04 mL, 0.36 mmol) and a catalytic amount of NaI were added and the mixture was warmed to r.t. and stirred for one hour. The reaction was then quenched with ice and extracted with Et$_2$O. The organic layer was dried (MgSO$_4$), filtered and concentrated. Purification by column chromatography (Hex:EtOAc = 50:1 to 20:1) yielded **366** as a colourless oil. (74 mg, 73%).

^{1}H NMR (400 MHz, CDCl$_3$): δ 7.40-7.23 (m, 5 H), 4.73 (d, J = 11.7 Hz, 1 H), 4.62 (d, 11.7 Hz, 1 H), 4.37 (ddd, J = 6.6, 5.7, 2.7 Hz, 1 H), 3.54 (dd, J = 9.1, 6.6 Hz, 1 H), 3.45 (dd, J = 8.6, 2.7 Hz, 1 H), 3.43 (dd, J = 9.1, 5.7 Hz, 1 H), 3.30 (s, 3 H), 2.92 (m, 1 H), 1.75 (d, J = 2.3 Hz, 3 H), 1.23 (d, 6.8 Hz, 3 H), 1.15 – 1.05 (m, 21 H).

^{13}C NMR (100 MHz, CDCl$_3$): δ 139.7 (C), 128.7 (CH), 128.1 (CH), 83.5 (CH$_2$), 82.6 (C), 77.6 (C), 74.8 (CH$_2$), 73.7 (CH$_2$), 72.9 (CH), 59.3 (CH$_3$), 27.7 (CH), 18.8 (CH$_3$), 18.5 (CH$_3$), 13.5 (CH), 4.1 (CH$_3$).

IR [cm^{-1}]: 3065, 3030, 2943, 2891, 2866, 1731, 1497, 1454, 1384, 1367, 1304, 1249, 1198, 1099, 1063, 1028.

[α]$_D^{20}$ -2.5 (c 0.3 CHCl$_3$).

R$_f$ (Hex/EtOAc = 10/1): 0.56

[(*1S,2S,3R*)-2-(4-methoxybenzyloxy)-1-(methoxymethyl)-3-methyl-4-hexynyl]oxy(triisopropyl)silane

Compound **387** (186 mg, 0.24 mmol) was dissolved in 5 mL of DMF, cooled to 0 °C and then NaH suspension (34 mg, 0.85 mmol, 60 %) was added. The mixture was stirred for 30 min at 0 °C, then PMBCl (177 mg, 1.13 mmol) and TBAI (18mg, 0.06 mmol) was added and the mixture was warmed to r.t. and stirred overnight. The reaction was then quenched with ice-cold 1M NaOH. After stirring for 15 min, the solution was extracted 4 times with Et$_2$O. The combined organic layers were washed with brine, dried over MgSO$_4$, filtered and concentrated under reduced pressure. Purification by column chromatography (Hex : EtOAc = 50:1 to 20:1) yielded **368** as a colourless oil. (180 mg, 71 %)

^{1}H NMR (400 MHz, CDCl$_3$): δ = 7.28 (d, J = 8.6 Hz, 2 H), 6.86 (d, J =8.6 Hz, 2 H), 4.64 (d, J = 11.2 Hz, 1 H), 4.54 (d, J = 11.2 Hz, 1 H), 4.34 (ddd, J = 6.5, 6.1, 2.7 Hz, 1 H), 3.79 (s, 3 H), 3.54 (dd, J = 9.1, 6.5 Hz, 1 H), 3.43 (m, 2 H), 3.31 (s, 3 H), 2.89 (m, 1 H), 1.75 (d, J =2.3 Hz, 3 H), 1.21 (d, J =7.1 Hz, 3 H), 1.08 (m, 21 H).

^{13}C NMR (100 MHz, CDCl$_3$): δ = 159.0 (C), 131.3 (C), 129.0 (CH), 113.6 (CH), 82.7 (CH), 82.1 (C), 77.1 (C), 74.3 (CH$_2$), 72.8 (CH$_2$), 72.4 (CH), 58.7 (CH$_3$), 55.3 (CH$_3$), 27.1 (CH), 18.2 (CH$_3$), 18.2 (CH$_3$), 18.0 (CH$_3$), 12.9 (CH), 3.6 (CH$_3$).

IR [cm^{-1}]: 2941, 2866, 1613, 1586, 1514, 1464, 1383, 1302, 1248, 1172, 1096, 1062, 1038.

MS (EI) m/z 405 (M$^+$-43, 4), 375 (0.3), 267 (1), 187 (4), 145 (4), 121 (100).

HRMS(EI) calcd. For C$_{23}$H$_{37}$O$_4$Si: 405.2461, Found: 405.2466.

[α]$_D^{20}$ -4.8 (c 1.9, CHCl$_3$).

R$_f$ (Hex/EtOAc = 10/1): 0.48

(4*S*,5*S*)-4-Methoxymethyl-2-(4-methoxy-phenyl)-5-((*R*)-1-methyl-but-2-ynyl)-1,3-dioxolane

Compound **387** (36 mg, 0.08 mmol) was dissolved in 2 mL of THF and TBAF in THF (0.24 mL, 0.24 mmol, 1 M) was added. After stirring at r.t. overnight, the reaction was quenched by addition of saturated NaHCO$_3$ solution and subsequently extracted 4 times with a total of 100 mL of ethyl acetate. The combined organic layers were washed with brine, dried over magnesium sulfate, filtered and the solvent removed under reduced pressure. Purification by column chromatography using Hex:EtOAc = 7:1 to 2:1 yielded the secondary alcohol as a colourless oil. (22 mg, 94 %).

The free secondary alcohol (22 mg, 0.08 mmol) was dissolved in 1 mL of DCM, 100 mg of powdered 4Å molecular sieves and DDQ (22 mg, 0.10 mmol) was added at 0 °C. After stirring at this temperature for 1.5 h, the reaction was quenched by addition of a 1:1 mixture of sat. NaHCO$_3$ and 1 M Na$_2$S$_2$O$_3$. The mixture was then extracted 2 times with DCM. The combined organic layers were washed with brine, dried over MgSO$_4$, filtered and concentrated under reduced pressure. Purification by column chromatography (Hex:EtOAc = 5:1) yielded **369** as a colourless oil. (12 mg, 55 %)

369a:

^{1}H NMR (400 MHz, CDCl$_3$): δ = 7.43 (d, *J* = 8.7 Hz, 1 H), 7.42 (d, *J* = 8.7 Hz, 0.3 H), 6.88 (d, *J* = 8.9 Hz, 0.3 H), 6.87 (d, *J* = 8.8 Hz, 1 H), 5.90 (s, 1 H), 5.86 (s, 0.3 H), 4.26 (ddd, *J* = 5.9, 5.8, 3.0 Hz, 1 H), 4.31 (ddd, *J* = 6.4, 6.4, 3.0 Hz, 0.3 H), 3.81 (dd, *J* = 7.8, 5.8 Hz, 1 H), 3.80 (s, 4 H), 3.75 (dd, *J* = 3.0, 0.4 Hz, 0.3 H), 3.73 (dd, *J* = 5.9, 2.6 Hz, 0.3 H), 3.68 (dd, *J* = 10.6, 3.0 Hz, 1 H), 3.61 (dd, *J* = 10.6, 5.8 Hz, 1 H), 3.60 (dd, *J* =

10.9, 6.3 Hz, 0.3 H), 3.45 (s, 3 H), 3.44 (s, 1 H), 2.68 (m, 1.3 H), 1.81 (d, J = 2.3 Hz, 3 H), 1.80 (d, J = 2.3 Hz, 0.3 H), 1.32-1.17 (m, 32 H).

^{13}C NMR (100 MHz, CDCl$_3$): δ = 160.9 ©, 130.15 (CH), 129.8 ©, 128.7 (CH), 114.1 (CH), 114.0 (CH), 104.1 (CH), 103.7 (CH), 81.6 (CH), 81.4 (CH), 80.9 (CH), 80.0 (CH), 79.9 ©, 79.7 ©, 78.8 ©, 78.4 ©, 74.5 ((CH2)), 74.2 ((CH2)), 59.8 ((CH3)), 55.70 (CH), 55.68 (CH), 31.1 (CH), 30.5 (CH), 18.7 ((CH3)), 18.6 ((CH3)).

R$_f$ (Hex/EtOAc = 3/1): 0.45

[(*1S,2S,3R*)-5-iodo-2-(4-methoxybenzyloxy)-1-(methoxymethyl)-3-methyl-4*E*-hexenyl]oxy(triisopropyl)silane

Cp$_2$ZrCl$_2$ (59 mg, 0.20 mmol) was dissolved in a Schlenk flask in anhydrous THF (1.5 mL) and cooled to 0 °C. DIBAL-H in hexanes (0.2 mL, 0.20 mmol, 1 M) was added and the resulting slurry was stirred at r.t. for 1.5 hours. The stirring was stopped, and after 5 minutes the supernatant liquid was removed with a syringe. To the white precipitate was then added a solution of **368** (30mg, 0.07 mmol) in freshly distilled benzene (2 mL). The mixture was then heated to 40 °C. After 5 minutes the solution became clear and stirring was continued at this temperature for 3 hours. Then, the oil bath was removed and the reaction mixture cooled to 0°C. A solution of iodine (51 mg, 0.20 mmol) in benzene (2 mL) was then added slowly. Immediately after completion of the addition, the reaction was quenched by addition of a 1M Na$_2$S$_2$O$_3$ solution (5 mL). When both layers were colourless, the organic layer was separated and the aqueous layer was extracted 4 times with diethyl ether. The combined organic layers were washed with brine, dried over MgSO$_4$, filtered and concentrated under reduced pressure.

Purification by column chromatography (Hex:EtOAc 50:1 to 20:1) yielded **410** as a colourless oil. (38 mg, 98 %)

^{1}H NMR (400 MHz, CDCl$_3$): δ = 7.26 (d, J =8.6 Hz, 2 H), 6.88 (d, J =8.6 Hz, 2 H), 6.13 (dq, J = 9.7, 1.5 Hz, 1 H), 4.61 (d, J = 11.1 Hz, 1 H), 4.47 (d, J = 11.1 Hz, 1 H), 4.07 (ddd, J = 6.2, 4.1, 4.1 Hz, 1 H), 3.80 (s, 3 H), 3.56 (dd, J = 9.5, 3.6 Hz, 1 H), 3.34 (m, 2 H), 3.31 (s, 3 H), 2.71 (m, 1 H), 2.36 (d, J =1.5 Hz, 3 H), 1.08 (m, 21 H), 1.00 (d, J =6.8 Hz, 3 H).

^{13}C NMR (100 MHz, CDCl$_3$): δ = 158.1 ©, 143.7 (CH), 129.8 ©, 128.3 (CH), 128.2 (CH), 112.7 (CH), 112.6 (CH),91.6 ©, 82.2 (CH), 73.7($_{(CH2)}$), 72.3 ($_{(CH2)}$), 71.8 (CH), 57.8 ($_{(CH3)}$), 54.3 ($_{(CH3)}$), 35.6 (CH), 26.7 ($_{(CH3)}$), 17.3 ($_{(CH3)}$), 15.5 ($_{(CH3)}$), 11.8 (CH).

IR [cm^{-1}]: 2962, 2866, 1612, 1586, 151, 1463, 1382, 1302, 1250, 1196, 1172, 1110.

MS (EI) m/z 533 (M$^+$-43, 3), 407 (1), 285 (2), 187 (6), 121 (100).

HRMS(EI) calcd. For C$_{23}$H$_{38}$O$_4$Isi: 533.1584, Found: 533.1598.

[α]$_D^{20}$ – 2.5 (c 1.3, CHCl$_3$).

R$_f$ (Hex/EtOAc = 10/1): 0.48

[(*1S,2S,3R*)-5-iodo-2-(benzyloxy)-1-(methoxymethyl)-3-methyl-4*E*-hexenyl]oxy(triisopropyl)silane

^{1}H NMR (400 MHz, CDCl$_3$): δ = 7.40-7.31 (m, 5 H, Ph), 6.16 (dq, J = 9.8, 1.4 Hz, 1 H), 4.69 (d, J = 11.4 Hz, 1 H), 4.56 (d, J = 11.4 Hz, 1 H), 4.10 (m, 1 H), 3.57 (dd, J = 9.6,

3.8 Hz, 1 H), 3.37 (dd, J = 9.6, 6.3 Hz, 1 H), 3.31 (s, 3 H), 2.74 (m, 1 H), 2.37 (d, J = 1.4 Hz, 3 H), 1.10 (m, 21 H), 1.03 (d, J = 6.9 Hz, 3 H).

^{13}C NMR (100 MHz, CDCl$_3$): δ 144.6 (CH), 138.7 (C), 128.3 (CH), 127.6 (CH), 127.3 (CH), 92.7 (C), 83.5 (CH$_2$), 74.7 (CH$_2$), 73.7 (CH$_2$), 72.7 (CH), 58.8 (CH$_3$), 36.6 (CH), 18.3 (CH$_3$), 18.1 (CH$_3$), 12.8 (CH).

12 APPENDIX

12.1 Used Abbreviations

Ac	Acetyl
AIBN	*N,N'*-(bis)-Azo-isobutyronitrile
APT	Attached Proton Test
BHT	2,6 bis-*tert*-butyl-4-methyl phenol
Bn	Benzyl
Bz	Benzoyl
CuTC	Copper-(I)-thiophen-2-carboxylate
dba	Dibenzylidenacetone
DBU	1,8-Diazabicyclo[5.4.0]undec-7-ene
DCM	Dichloromethane
DDQ	2,3-Dichloro-5,6-dicyano-quinone
DHP	4,5-Dihydropyrane
DIAD	Diisopropylazodicarboxylate
DIBAl-H	Diisobutylaluminium hydride
DIC	Diisopropylcarbodiimide
DIPT	Diisopropyltartrate
DMAP	4-Dimethylamino-pyridine
DMDO	Dimethyldioxirane
DMP	Dess-Martin periodinane
DOSY	Diffusion-ordered spectroscopy
dppf	(bis)-diphenylphospine-1,1'-ferrocene
EDC·HCl	1-Ethyl-3-(3-dimethylaminopropyl) carbodiimide hydrochloride)
Et	Ethyl
EtOAc	Ethyl acetate
Hex	Hexanes
HMPA	Hexamethylphosphoramide
HTS	High-throughput screening
IBX	2-Iodoxybenzoic acid
IMDA	Intramolecular Diels-Alder
KHMDS	Potassium hexamethyldisilazide
LDA	Lithium diisopropylamide

LHMDS	Lithium hexamethyldisilazide
*m*CPBA	*meta*-chloroperbenzoic acid
Me	Methyl
MIC	Minimal inhibitory concentration
MOM	Methoxymethyl
Ms	Mesyl (Methylsulfonate)
NaHMDS	Sodium hexamethyldisilazide
NMO	*N*-Methyl-*N*-morpholine oxide
NMP	*N*-Methyl pyrrolidinone
Piv	Pivaloate
PMB	*para*-Methoxy-benzyl
PMP	*para*-Methoxy-phenyl
psi	Pounds per square inch
p-TsOH	*para*-Toluene sulfonic acid
pyr	Pyridine
r.t.	Room temperature
RCM	Ring closing metathesis
Rf	Retention factor
SAR	Structure-activity relationship
sat.	saturated
SEM	2-(Trimethylsilyl)ethoxymethyl
SET	Single-electron transfer
TADA	Transannular Diels-Alder
TBAF	Tetrabutylammonium fluoride
TBAI	Tetrabutylammonium iodide
TBDPS	*tert*-Butyldiphenylsilyl
TBHP	*tert*-Butylhydroperoxide
TBS	*tert*-Butyldimethylsilyl
TES	Triethylsilyl
Tf	Triflat (Trifluorosulfonate)
THF	Tetrahydrofurane
THP	Tetrahydropyrane
TIPS	Triisopropylsilyl
TLC	Thin layer chromatography
TMS	Trimethylsilyl
TPAP	Tetrapropylammonium perruthenate

12.2 Single-crystal diffraction data

(2R,3aR,5aR,7aS,10aS,10bS,10cR)-2-methoxy-2,3,3a,7a,8,10a,10b,10c-octahydro-5aH-1,9-dioxa-dicyclopenta[a,h]naphthalen-10-one (**225**)	
Empirical formula	$C_{15}H_{18}O_4$
Formula weight	262.29
Space group	$P2_12_12_1$
Cell symmetry	orthorhombic
Temperature (K)	100(2)
a (Å)	6.3696(2)
b (Å)	7.2069(2)
c (Å)	29.1029(10)
V (Å^3)	1335.97(7)
Z	4
Density (g cm^{-3})	1.304
λ (Å)	0.71073
Theta range for data collection (°)	2.80 to 28.34
hkl (min/max)	-8<=h=>8, -9<=k=>9, -38<=l=>38
F(000)	560
Total number of reflections	43529
Number of unique reflections	1969
R (int)	0.0751
Absorption cooefficient (mm^{-1})	0.094
Refinement method	Full-matrix least-squares on F^2
R1	0.0388
$wR2$-value	0.1003

Atomic coordinates (x 10⁴) and equivalent isotropic displacement parameters (Å x 10³)				
Atom	x	y	z	U(eq)
O(1)	8286(3)	1369(2)	7833(1)	29(1)
O(2)	6795(2)	1937(2)	8555(1)	21(1)
O(3)	8569(2)	5202(2)	10135(1)	21(1)
O(4)	5301(2)	4114(2)	10098(1)	22(1)
C(1)	6115(3)	3823(3)	8598(1)	17(1)
C(2)	4147(3)	3885(3)	8304(1)	21(1)
C(4)	6420(4)	1409(3)	8087(1)	22(1)
C(3)	4949(4)	2881(3)	7876(1)	26(1)
C(5)	9640(4)	-112(4)	7973(1)	38(1)
C(6)	5636(3)	4413(3)	9086(1)	15(1)
C(7)	4950(3)	6480(3)	9066(1)	17(1)
C(8)	3674(3)	6954(3)	8639(1)	22(1)
C(9)	3326(3)	5830(3)	8287(1)	24(1)
C(10)	7499(3)	4113(3)	9409(1)	15(1)
C(12)	10344(3)	5415(3)	9825(1)	22(1)
C(11)	9389(3)	5402(3)	9340(1)	18(1)
C(14)	6931(3)	4443(3)	9909(1)	17(1)

Appendix

Bond Lengths (Å)	
O(1)-C(4)	1.401(3)
O(1)-C(5)	1.432(3)
O(2)-C(1)	1.432(2)
O(2)-C(4)	1.434(2)
O(3)-C(14)	1.350(2)
O(3)-C(12)	1.455(3)
O(4)-C(14)	1.198(3)
C(1)-C(6)	1.513(3)
C(1)-C(2)	1.518(3)
C(2)-C(9)	1.497(3)
C(2)-C(3)	1.527(3)
C(4)-C(3)	1.543(3)
C(6)-C(10)	1.529(3)
C(6)-C(7)	1.553(3)
C(7)-C(15)	1.507(3)
C(7)-C(8)	1.524(3)
C(8)-C(9)	1.324(3)
C(10)-C(14)	1.519(3)
C(10)-C(11)	1.534(3)
C(12)-C(11)	1.537(3)
C(11)-C(16)	1.499(3)
C(15)-C(16)	1.329(3)

Bond Angles (°)	
C(4)-O(1)-C(5)	112.06(18)
C(1)-O(2)-C(4)	106.52(15)
C(14)-O(3)-C(12)	109.90(14)
O(2)-C(1)-C(6)	114.19(16)
O(2)-C(1)-C(2)	103.21(16)
C(6)-C(1)-C(2)	110.73(16)
C(9)-C(2)-C(1)	109.54(17)
C(9)-C(2)-C(3)	122.26(18)
C(1)-C(2)-C(3)	99.74(16)
O(1)-C(4)-O(2)	111.48(17)
O(1)-C(4)-C(3)	108.62(17)
O(2)-C(4)-C(3)	107.23(16)
C(2)-C(3)-C(4)	101.82(16)
C(1)-C(6)-C(10)	112.32(16)
C(1)-C(6)-C(7)	106.94(15)
C(10)-C(6)-C(7)	112.13(15)
C(15)-C(7)-C(8)	108.64(17)
C(15)-C(7)-C(6)	112.63(16)
C(8)-C(7)-C(6)	113.28(16)
C(9)-C(8)-C(7)	125.65(19)
C(8)-C(9)-C(2)	119.28(18)
C(14)-C(10)-C(6)	112.41(16)
C(14)-C(10)-C(11)	102.53(15)
C(6)-C(10)-C(11)	116.34(16)
O(3)-C(12)-C(11)	105.17(16)
C(16)-C(11)-C(10)	112.14(16)
C(16)-C(11)-C(12)	110.24(17)
C(10)-C(11)-C(12)	101.24(15)

O(4)-C(14)-O(3)	121.68(17)
O(4)-C(14)-C(10)	128.01(18)
O(3)-C(14)-C(10)	110.31(16)
C(16)-C(15)-C(7)	125.27(18)
C(15)-C(16)-C(11)	123.99(19)

12.3 Literature

[1] Walsh, C. T.; Wright, G. *Chem. Rev.* **2005**, *105*, 391 – 393.

[2] Nussbaum, F.; Brands, M.; Hinzen, B.; Weigand, S.; Häbich, D. *Angew. Chem. Int. Ed.* **2006**, *45*, 5072 – 5129; *Angew. Chem.* **2006**, *118*, 5194 – 5254.

[3] Fleming, A. *Br. J. Exp. Pathol.* **1929**, *10*, 226–236.

[4] For a review on the syntheses of misassigned natural products: Nicolaou, K. C.; Snyder, S. A. *Angew. Chem. Int. Ed.* **2005**, *44*, 1012-1044; *Angew. Chem.* **2005**, *117*, 1036 – 1069.

[5] a) Woodward, R. B. et al. *J. Am. Chem. Soc.* **1981**, *103*, 3210 – 3213; b) Woodward, R. B. et al. *J. Am. Chem. Soc.* **1981**, *103*, 3213 – 3215; c) Woodward, R. B. et al. *J. Am. Chem. Soc.* **1981**, *103*, 3215 – 3217.

[6] Moellering, R. C. Jr. *Clin. Infect. Dis.* **2006**, *42*, S3–S4.

[7] Nicolaou, K. C.; Mitchell, H. J.; Jain, N. F.; Winssinger, N.; Hughes, R.; Bando, T. *Angew. Chem. Int. Ed.* **1999**, *38*, 240 – 244; *Angew. Chem.* **1999**, *111*, 253 – 255.

[8] Levy, S. B. *Adv. Drug Deliv. Rev.* **2005**, *57*, 1446 – 1450.

[9] a) Hiramatsu, K.; Okuma, K.; Ma, X. X.; Yamamoto, M.; Hori, S.; Kapi, M. *Curr. Opin. Infect. Dis.* **2002**, *15*, 407 – 413; b) Weigel, L. M.; Clewell, D. B.; Gill, S. R.; Clark, N. C.; McGougal, L. K.; Flannagan, S. E.; Kolonay, J. F.; Shetty, J.; Killgore, G. E.; Tenover, F. C. *Science* **2003**, *302*, 1569 – 1571.

[10] Speitling, M. *Dissertation*, Universität Göttingen, **1998**.

[11] Celmer, W. D.; Cullen, W. P.; Moppett, C. E.; Jefferson, M. T.; Huang, L. H.; Shibakawa, R.; Tone, J. *United States Patent* US 4,418,883, **1979**.

[12] Celmer, W. D.; Chmurny, G. N.; Moppett, C. E.; Ware, R. S.; Watts, P. C.; Whipple, E. B. *J. Am. Chem. Soc.* **1981**, *102*, 4203 – 4209.

[13] Cane, D. E.; Yang, C.-C. *J. Am. Chem. Soc.* **1984**, *106*, 784-787.

[14] Magerlein, B. J.; Reid, R. J. *J. Antibiot.* **1982**, *35*, 254-255

[15] Whaley, H. A.; Chidester, C. G.; Mizsak, S. A.; Wnuk, R. J. *Tetrahedron Lett.* **1980**, *21*, 3659-3662

[16] Celmer, W. D.; Cullen, W. P.; Moppett, C. E.; Jefferson, M. T.; Huang, L. H.; Shibakawa, R.; Tone, J. *United States Patent* US 4,224,314, **1980**.

[17] Celmer, W. D.; Cullen, L. H.; Shibakawa, R.; Tone, J. *United States Patent* US 4,436,747, **1984**.

[19] Ōmura, S.; Iwata, R.; Iwai, Y.; Taga, S.; Tanake, Y.; Tomoda, H. *J. Antibiot.* **1985**, *38*, 1322 - 1326

[20] Jackson, M.; Karwowski, J. P.; Theriault, R. J.; Fernandes, P. B.; Semon, R. C. *J. Antibiot.* **1987**, *40*, 1375 - 1381

[21] Rasmussen, R. R.; Scherr, M. H.; Whittern, D. N.; Buko, A. M.; McAlpine, J. B. *J. Antibiot.* **1987**, *40*, 1383 - 1393

[22] Gouda, H.; Sunazuka, T.; Ui, H.; Handa, M.; Sakoh, Y.; Iwai, Y.; Hirono, S.; Ōmura, S. *Proc. Natl. Acad. Sci. U.S.A.* **2005**, *102*, 18286-18291.

[23] Tomoda, H.; Iwata, R.; Takahashi, Y.; Iwai, Y.; Ōiwa, R.; Ōmura, S.; *J. Antibiot.* **1986**, *39*, 1205 – 1210.

[24] Handa, M.; Ui, H.; Yamamoto, D.; Monma, S.; Iwai, Y.; Sunazuka, T.; Omura, S. *Heterocycles* **2003**, *59*, 497-500.

[25] Snyder, W. C.; Rinehart, K. L. *J. Am. Chem. Soc.* **1984**, *106*, 787-789.

[26] Cane, D. E.; Yang, C.-C. *J. Antibiot.* **1985**, *38*, 423-426.

[27] Cane. D. E.; Ott, W. R. *J. Am. Chem. Soc.* **1988**, *110*, 4840-4841; Cane, D. E.; Tan, W. T.; Ott, W. R. *J. Am. Chem. Soc.* **1993**, *115*, 527-535.

[28] Cane, D. E.; Luo, G. *J. Am. Chem. Soc.* **1995**, *117*, 6633-6634.

[29] For reviews on biosynthetic Diels-Alder reactions, see: a) Williams, R. M.; Stocking, E. M. *Angew. Chem. Int. Ed.* **2003**, *42*, 3078-3115; b) Oikawa, H.; Tokiwano, T. *Nat. Prod. Rep.* **2004**, *21*, 321-352.

[30] McAlpine, J. B.; Mitscher, L. A.; Jackson, M.; Rasmussen, R. R.; van der Velde, D.; Veliz, E.; *Tetrahedron* **1996**, *52*, 10327-10334.

[31] Kallmerten, J. *Tetrahedron Lett.* **1984**, *25*, 2843-2846.

[32] Plata, D. J.; Kallmerten, J. *J. Am. Chem. Soc.* **1988**, *110*, 4041 – 4042.

[33] Corey, E. J.; Trybulski, E. J.; Melvin, L. S.; Nicolaou, K. C.; Secrist, J. A.; Lett, R.; Sheldrake, P. W.; Falck, J. R.; Brunelle, D. J.; Haslanger, M. F.; Kim, S.; Yoo, S. *J. Am. Chem. Soc.* **1978**, *100*, 4618.

[34] a) Vedejs, E. *J. Am. Chem. Soc.* **1974**, *96*, 5966; b) Vedejs, E.; Telschow, J. E. *J. Org. Chem.* **1976**, *41*, 740.

[35] Corey, E. J.; Nicolaou, K. C. *J. Am. Chem. Soc.* **1974**, *96*, 5614-5616.

[36] Wittman, M. D.; Kallmerten, J. *J. Org. Chem.* **1988**, *53*, 4631-4633.

[37] Rossano, L. T.; Plata, D. J.; Kallmerten, J. *J. Org. Chem.* **1988**, *53*, 5189-5191

[38] Rossano, L. T. *Ph. D Thesis*, Syracuse University, **1990**.

[39] Jones, R. F.; Tunnicliffe, J. H. *Tetrahedron Lett.* **1985**, *26*, 5845-5848.

[40] a) Roush, W. R.; Coe, J. W. *Tetrahedron Lett.* **1987**, *28*, 931-934; b) Coe, J. W.; Roush, W. R. *J. Org. Chem.* **1989**, *54*, 915-930.

[41] a) Roush, W. R.; Halterman, R. L. *J. Am. Chem. Soc.* **1986**, *108*, 294-296; b) Roush, W. R.; Hoong, J. L. K.; Palmer, M. A. J.; Park, J. C. *J. Org. Chem.* **1990**, *55*, 4109-4117; c) Roush, W. R.; Hoong, J. L. K.; Palmer, M. A. J.; Straub, J. A.; Palkowitz, A. D. *J. Org. Chem.* **1990**, *55*, 4117-4126.

[42] Roush, W. R.; Koyama, K.; Curtin, M. L.; Moriarty, K. J. *J. Am. Chem. Soc.* **1996**, *118*, 7502-7512.

[43] a) Gössinger, E.; Graupe, M.; Zimmermann, K. *Monatsh. Chem.* **1993**, *124*, 965-979; b) Gössinger, E.; Graupe, M.; Kratky, C.; Zimmermann, K. *Tetrahedron* **1997**, *53*, 3083-3100.

[44] Auer, E.; Gössinger, E.; Graupe, M. *Tetrahedron Lett.* **1997**, 6577-6580.

[45] a) Enev, V. S.; Drescher, M.; Kählig, H.; Mulzer, J. *Synlett*, **2005**, *14*, 2227 – 2229; b) Mulzer, J.; Castagnolo, D.; Felzmann, W.; Marchart, S.; Pilger, C.; Enev, V. S.; *Chem. Eur. J.* **2006**, *12*, 5992-6001.

[46] Murray, L. M.; O'Brien, P.; Taylor, R. J. K. *Org. Lett.* **2003**, *5*, 1943.

[47] Marchart, S.; Mulzer, J.; Enev, V. S. *Org. Lett.* **2007**, ASAP.

[48] Raveendranath, P. C.; Blazis, V. J.; Agyei-Aye, K.; Hebbler, A. K.; Gentile, L. N.; Hawkins, E. S.; Johnson, S. C.; Baker, D. C. *Carbohydr. Res.* **1994**, *253*, 207-223.

[49] Barton, D. H. R.; Liu, W. *Tetrahedron*, **1997**, *53*, 12067-12088.

[50] Siozaki, M.; Arai, M.; Kobayashi, Y.; Kasuya, A.; Miyamoto, S. *J. Org. Chem.* **1994**, *59*, 4450-4460.

[51] Takai, K.; Nitta, K.; Utimoto, K. *J. Am. Chem. Soc.* **1986**, *108*, 7408-7410.

[52] Rodriguez, J. B. *Tetrahedron* **1999**, *55*, 2157-2170.

[53] a) For a review on diimide reductions, see: Pasto, D. J.; Taylor, R. T. *Org. React.* **1991**, *40*, 91-155; b) Hoffmann, R. W.; Hense, A. *Liebigs Ann.* **1996**, 1283-1288.

[54] Sonogashira, K.; Tohda, Y.; Hagihara, N. *Tetrahedron Lett.* **1975**, *16*, 4467-4470.

[55] Ditrich, K.; Hoffmann, R. W. *Tetrahedron Lett.* **1985**, *26*, 6325-6328.

[56] Andrews, G. C.; Crawford, T. C.; Bacon, B. E. *J. Org. Chem.* **1981**, *46*, 2976-2977.

[57] Hubschwerlen, C. *Synthesis* **1986**, 962-964

[58] Fazio, F.; Schneider, M. P. *Tetrahedron: Asymmetry* **2000**, *11*, 1869-1876.

[59] Takano, S.; Kurotaki, A.; Takahashi, M.; Ogasawara, K. *Synthesis* **1986**, 403-406.

[60] Nilsson, K.; Ullenius, C. *Tetrahedron* **1994**, *50*, 13173-13180.

[61] For a review on reactions for this purpose, see: Rezgui, F.; Amrib, H.; El Gaïed, M. M. *Tetrahedron* **2003**, *59*, 1369–1380.

[62] Rezgui, F.; El Gaïed, M. M. *Tetrahedron Lett.* **1998**, *39*, 5965-5966.

[63] Takano, S.; Tanaka, M.; Seo, K.; Dirama, M.; Ogasawara, K. *J. Org. Chem.* **1985**, *50*, 931-936.

[64] Mori, K.; Khlebnikov, V. *Liebigs Ann. Chem.* **1993**, 77-82.

[65] Calderon, A.; de March, P.; El Arrad, M.; Font, J. *Tetrahedron* **1994**, *50*, 4201-4214.

[66] Tamura, R.; Saegusa, K.; Kakihana, M.; Oda, D. *J. Org. Chem.* **1988**, *53*, 2723-2728.

[67] Stork, G.; Zhao, K. *Tetrahedron Lett.* **1989**, *30*, 2173-2174.

[68] Jung, M. D.; Light, L. A. *Tetrahedron Lett.* **1982**, *23*, 3851-3854.

[69] a) Stille, J. K.; Simpson, J. H. *J. Am. Chem. Soc.* **1987**, *109*, 2138-2152. b) Stille, J. K. *Angew. Chem. Int. Ed.* **1986**, *25*, 508-524; *Angew. Chem.* **1986**, *98*, 504-519.

[70] Collington, E. W.; Meyers, A. I. *J. Org. Chem.* **1971**, *36*, 3044-3045.

[71] Siegel, K.; Brückner, R. *Chem. Eur. J.* **1998**, *4*, 1116-1122.

[72] Franci, X.; Martina, S. L. X.; McGrady, J. E.; Webb, M. R.; Donald, C.; Taylor, R. J. K. *Tetrahedron Lett.* **2003**, *44* 7735-7740.

[73] Still, W. C.; Gennari, C. *Tetrahedron Lett.* **1983**, *24*, 4405–4408

[74] Ando, K. *J. Org. Chem* **1999**, 6815-6821

[75] Munakata, R.; Katakai, H.; Ueki, T.; Kurosaka, J.; Takao, K.-i.; Tadano, K.-i. *J. Am. Chem. Soc.* **2003**, *125*, 14722 – 14723.

[76] Kadota, I.; Takamura, H.; Sato, K.; Ohno, A.; Matsuda, K.; Satake, M.; Yamamoto, Y. *J. Am. Chem. Soc.* **2003**, *125*, 11893 – 11899.

[77] For a review on DMSO oxidations, see: Tidwell, T.T. *Synthesis* **1990**, 857-870.

[78] a) Barriault, L.; Thomas, J. D. O.; Clément, R. *J. Org. Chem.* **2003**, *68*, 2317-2323; b) Sieburth, S.; Fensterbamk, L. *J. Org. Chem.* **1992**, *57*, 5279-5281; c) Stork, G.; Chan, T. Y.; Breault, G. A. *J. Am. Chem. Soc.* **1992**, *114*, 7578-7579; d) Stork, G.; Chan, T. Y. *J. Am. Chem. Soc.* **1995**, *117*, 6595-6596; e) Olsson, R.; Bertozzi, F.; Fredj, T. *Org. Lett.* **2000**, *2*, 1283-1286; f) Batey, R. A.; Thadani, A. N.; Lough, A. J. *J. Am. Chem. Soc.* **1999**, *121*, 450-451 and references cited therein.

[79] Ireland, R. E.; Norbeck, D. W. *J. Org. Chem.* **1985**, *50*, 2198-2200.

[80] Bressette, A. R.; Glover, L. C. *Synlett*, **2004**, 738-740.

[81] a) Blackburn, L.; Xudong, W.; Taylor, R. J. K. *Chem. Commun.* **1999**, 1337-1338; b) For a review on this matter, see: Taylor, R. J. K.; Reid, M.; Foot, J.; Raw, S. A. *Acc. Chem. Res.* **2005**, *38*, 851-869.

[82] a) Huang, C. C. *J. Labelled Compd. Radiopharm.* **1987**, *24*, 676-681 b) Barrett, A. G. M.; Hamprecht, D.; Ohkubo, M. *J. Org. Chem.* **1997**, *62*, 9376-9378.

[83] Cayzer, T. N.; Paddon-Row, M. N.; Sherburn, M. S. *Eur. J. Org. Chem.* **2003**, 4059-4068.

[84] For a review, see: Matsumoto, K.; Hamana, H.; Iida, H. *Helv. Chim. Acta* **2005**, *88*, 2033-2234.

[86]   Suzuki, T.; Usui, K.; Miyake, Y.; Namikoshi, M.; Nakada, M. *Org. Lett.* **2004**, *6*, 553-556.

[87]   Nicolaou, K. C.; Snyder, S. A.; Huang, X.; Simonsen, K. B.; Koumbis, A. E.; Bigot, A. *J. Am. Chem. Soc.* **2004**, *126*, 10162 – 10173.

[88]   For reviews on this matter, see: a) Duncton, M. A. J.; Pattenden, G. *J. Chem. Soc., Perkin Trans. I* **1999**, 1235-1236; b) Nicolaou, K. C.; Bulger, P. G.; Sarlah, D. *Angew. Chem. Int. Ed.* **2005**, *44*, 4442 – 4489.

[89]   More, J. D.; Finney, N. S. *Org. Lett.* **2002**, *4*, 3001-3003.

[90]   Masamune, S.; Roush, W. R.; Sakai, T. *Tetrahedron Lett.* **1984**, *25*, 2183-2186.

[91]   Hodgson, D. M.; Foley, A. M.; Boutlon, L. T.; Lovell, P. J.; Maw, G. N. *J. Chem. Soc., Perkin Trans. I* **1999**, 2911-2922.

[92]   Brody, M. S.; Finn, M. G. *Tetrahedron Lett.* **1999**, *40*, 415-418.

[93]   a) Allred, G. D.; Liebeskind, L. S. *J. Am. Chem. Soc.* **1996**, *118*, 2748-2749; b) Paterson, I.; Man, J. *Tetrahedron Lett.* **1997**, *38*, 695-698.

[94]   Nicolaou, K. C.; Chakraborty, T. ; Piscopio, A. D.; Minowa, N, Bertinato, P. *J. Am. Chem. Soc.* **1993**, *115*, 4419-4420.

[95]   a) Renaldo, A. F.; Labadie, J. W.; Stille J. K. *Org. Synth.* **67**, 86-97; b) Bottaro, J. C.; Hanson, R. N.; Seitz, D. E. *J. Org. Chem.* **1981**, *46*, 5221-5222.

[96]   Ma, S.; Lu, X.; Li, Z. *J. Org. Chem.* **1992**, *57*, 709 – 713.

[97]   Stille, J. K. *J. Am. Chem. Soc.* **1986**, *108*, 3033-3040.

[98]   Colberg, J. C.; Rane, A.; Vaquer, J.; Soderquist, J. A. *J. Am. Chem. Soc.* **1993**, *115*, 6065-6071.

[99]   a) Zweifel, G.; Polston, N. L.; Whitney, C. C. *J. Am. Chem. Soc.* **1968**, *90*, 6243-6245; b) Ichikawa, J.; Moriya, T.; Sonoda, T.; Kobayashi, H. *Chem. Lett.* **1991**, 961-964.

[100] Takagi, J.; Takahashi, K.; Ishiyama, T.; Miyaura, N. *J. Am. Chem. Soc.* **2002**, *124*, 8001-8006.

[101] Shindo, M.; Sugioka, T.; Umaba, Y.; Shishido, K. *Tetrahedron Lett.* **2004**, *45*, 8863-8866.

[102] Dineen, T. A.; Roush, W. R.; *Org. Lett.* **2004**, *6*, 2043-2046.

[103] Zhang, S.; Zhang, D.; Liebeskind, L. S. *J. Org. Chem.* **1997**, *62*, 2312 – 2313.

[104] Lee, P. H.; Seomoon, D.; Lee, K. *Org. Lett.* **2005**, *7*, 343 – 345.

[105] Mukaiyama, T.; Izumi, J.; Shiina, I. *Chem. Lett.* **1997**, 187.

[106] a) for a review on the use of alkenyl zirconocenes, see: Wipf, P.; Kendall, C. *Chem. Eur. J.* **2002**, *8*, 1778-1784; b) Panek, J. S.; Hu, T. *J. Org. Chem.* **1997**, *62*, 4912-4913; C) Lee, T. W.; Corey, E. J. *J. Am. Chem. Soc.* **2001**, *123*, 1872-1877.

[107] a) Paterson, I.; Schlapbach, A. *Synlett* **1995**, 498-500; b) Paterson, I.; Florence, G. C.; Gerlach, K.; Scott, J. P.; Sereinig, N. *J. Am. Chem. Soc.* **2001**, *123*, 9535-9544.

[108] Andringa, H.; Heus-Kloos, Y. A.; Brandsma, L. *J. Organomet. Chem.* **1987**, *336*, C41-43.

[109] a) Jeffery, T. *Tetrahedron* **1996**, *52*, 10113-10130; b) Bhatt, U.; Christmann, M.; Quitschalle, M.; Claus, E.; Kalesse, M. *J. Org. Chem.* **2001**, *66*, 1885-1893.

[110] Constantieux, T.; Rodriguez, J. *Science of Synthesis,* Vol. 26, **2005** 413-462.

[111] Prantz, K. *Diplomarbeit*, Universität Wien, **2005**.

[112] Nakagawa, M.; Saegusa, J.; Tonozuka, M.; Obi, M.; Ciuchi, M.; Hino, T.; Ban, Y. *Org. Synth.* **56**, 49.

[113] For reviews on this topic, see: a) Ocampo, R.; Dolbier, W. R. Jr. *Tetrahedron* **2004**, *60*, 9325-9374; b) Fürstner, A. *Synthesis* **1989**, 571-590.

[114] Wessjohann, L. A.; Scheid, G. *Synthesis*, **1999**, 1-36.

[115] a) Obringer, M.; Colobert, F.; Neugnot, B.; Solladié, G. *Org. Lett.* **2003**, *5*, 629-632; b) Orsini, F.; Sello, G.; Manzo, A. M.; Lucci, E. M. *Tetrahedron: Asymmetry* **2005**, *16*, 1913-1918.

[116] Harada, T.; Mukaiyama, T. *Chem. Lett.* **1982**, 161-164.

[117] a) Rice, L. E.; Boston, C. M.; Finklea, H. O.; Suder, B. S.; Frazier, J. O.; Hudlicky, T. *J. Org. Chem.* **1984**, *49*, 1845-1848; b) Hudlicky, T.; Natchus, M. G.; Kwart, L. D.; Colwell, B. L. *J. Org. Chem.* **1985**, *50*, 4300-4306; c) Gurjar, M. K.; Reddy, D. S.; Bhadbhade, M. M.; Gonnade, R. G. *Tetrahedron*, **2004**, *60*, 10269-10275.

[118] Colvin, E. W.; Hamill, B. J. *J. Chem. Soc., Chem. Commun.* **1973**, 151-152.

[119] Sorg, A.; Brückner, R. *Synlett* **2005**, 289-293

[120] Blakemore, P. R. *J. Chem. Soc., Perkin Trans. 1* **2002**, 2563-2585.

[121] Truce, W. E.; Kreider, E. M.; Brand, W. W. *Org. React.* **1970**, *18*, 99.

[122] a) Nicolaou, K. C.; Baran, P. S.; Zhong, Y.-L.; Barluenga, S.; Hunt, K. W.; Kranich, R.; Vega, J. A. *J. Am. Chem. Soc.* **2002**, *124*, 2233-2244; b) Nicolaou, K. C.; Montagnon, T.; Baran, P. S.; Zhong, Y.-L. *J. Am. Chem. Soc.* **2002**, *124*, 2245-2258.

[123] Kruppa, A. I.; Taraban, M. B.; Shokhirev, N. V.; Svarovsky, S. A.; Leshina, T. V. *Chem. Phys. Lett.* **1996**, *285*, 316-322.

[124] MacCoss, R. N.; Balskus, E. P.; Ley, S. V. *Tetrahedron Lett.* **2003**, *44*, 7779-7781.

[125] Mee, S. P. H.; Lee, V.; Baldwin, J. E. *Angew. Chem. Int. Ed.* **2004**, *43*, 1132-1136.

[126] Farina, V.; Krishnan, B. *J. Am. Chem. Soc.* **1991**, *113*, 9585-9595.

[127] Srogl, J.; Allred, G. D.; Liebeskind, L. S. *J. Am. Chem. Soc.* **1997**, *119*, 12376-12377.

[128] Garg, N. K.; Hiebert, S.; Overman, L. E. *Angew. Chem. Int. Ed.* **2006**, *45*, 2912-2915.

[129] Vaz, B.; Dominguez, M.; Alvarez, R.; de Lera, A. R. *J. Org. Chem.* **2006**, *71*, 5914-5920.

[130] a) Corey, E. J.; Ensley, H. E. *J. Am. Chem. Soc.* **1975**, *97*, 6908-6909; b) Takeda, K.; Shibata, Y.; Sagawa, Y.; Urahata, M.; Funaki, K.; Hori, K.; Sasahara, H.; Yoshii, E. *J. Org. Chem.* **1985**, *50*, 4673-4681; c) Rath, J.-P.; Kinast, S.; Maier, M. E. *Org. Lett.* **2005**, *7*, 3089-3092.

[131] Williams, J. M.; Jobson, R. B.; Yasuda, N.; Marchesini, G.; Dolling, U.-H.; Grabowski, E. J. J. *Tetrahedron Lett.* **1995**, *36*, 5461-5464.

[132] Mulzer, J.; Giester, G.; Gilbert, M. *Helv. Chim. Acta* **2005**, *88*, 1560-1579.

[133] For a review, see: Hoveyda, A. H.; Evans, D. A.; Fu, G. C. *Chem. Rev.* **1993**, *93*, 1307 – 1370.

[134] Sharpless, K. B.; Michaelson, R. C. *J. Am. Chem. Soc.* **1973**, *95*, 6136 – 6137.

[135] Gadwood, R. C.; Lett, R. M.; Wissinger, J. E. *J. Am. Chem. Soc.* **1986**, *108*, 6343 – 6350.

[136] Rosenbeiger, D. *Diplomarbeit* Universität Wien, **2004**.

[137] Roush, W. R.; Hoong, L. K.; Palmer, M. A. J.; Straub, J. A.; Palkowitz, A. D. *J. Org. Chem.* **1990**, *55*, 4117-4126.

[138] Corey, E. J.; Fuchs, P. L. *Tetrahedron Lett.* **1972**, *36*, 3769-3772.

[139] Benechie, M; Skrydstrup, T.; Khuong-Huu, F. *Tetrahedron Lett.* **1991**, *32*, 7535-7538.

[140] a) Marshall, J.A. *Chem Rev.* **1996**, *96*, 31; b) Marshall, J.A.; Wang, X. *J .Org. Chem.* **1992**, *57*, 1242-1252. c) Marshall, J.A.; Chobanian, H. *Organic Syntheses* **2005**, *82*, 43-50.

[141] a) Buckle, M. J. C.; Fleming, I. *Tetrahedron Lett.* **1993**, *34*, 2383-2386. b) Marshall, J. A.; Maxson, K. *J. Org. Chem.* **2000**, *65*, 630-633. c) Buckle, M. J. C., Fleming, I.; Gil, S.; Pang, K. L. C. *Org. Biomol. Chem.* **2004**, *2*, 749-769.

[142] Rao, A. V. R.; Reddy, E. R.; Joshi, B. V.; Yadav, J. S. *Tetrahedron Lett.* **1987**, *28*, 6497-6500.

[143] Alcaraz, L.; Harnett, J. J.; Mioskowski, C.; Martel, J. P.; Le Gall, T.; Shin, D.-S.; Falck, J. R. *Tetrahedron Lett.* **1994**, *35*, 5449-5452.

[144] Bahadoor, A. B.; Flyer, A.; Micalizio, G. C. *J .Am. Chem. Soc.* **2005**, *127*, 3694-3695.

[145] a) Danheiser, R. L.; Carini, D. J.; Kwasigroch, C. A. *J. Org. Chem.* **1986**, *51*, 3870-3878. b) Danheiser, R. L.; Kwasigroch, C. A.; Tsai, Y.-M. *J. Am. Chem.Soc.* **1985**, *107*, 7233-7235.

[146] For a review on this topic, see: Masse, C. E.; Panek, J. S. *Chem. Rev.* **1995**, *95*, 1293-1316.

[147] Reetz, M. T. *Acc. Chem. Res.* **1993**, *26*, 462 – 468.

[148] Keck, G. E.; Castellino, S. *J. Am. Chem. Soc.* **1986**, *108*, 3847-3849.

[149] For ^{13}C-studies on BF$_3$-carbonyl complexes, see: a) Hartman, J. S.; Stilbs, P. *Tetrahedron Lett.* **1975**, *16*, 3497 – 3500; b) Torri, J.; Azzaro, M. *Bull. Soc. Chim. Fr.* **1978**, 283 – 291.

[151] For other protecting group induced transition state changes in related transformation, see: a) Mikami, K.; Matsukawa, S.; Sawa, E.; Harada, A.; Koga, N. *Tetrahedron Lett.* **1997**, *38*, 1951 – 1954; b) Figueras, S.; Martín, R.; Romea, P.; Urpí, F.; Vilarrasa, J. *Tetrahedron Lett.* **1997**, *38*, 1637 – 1640.

[152] For examples where these rules are followed see: a) Williams, D. R.; Wultz, M. W. *J. Am. Chem. Soc.* **2005**, *127*, 14550-14551. b) Reymond, S.; Cossy, J. *Eur. J. Org. Chem.* **2006**, 4800-4804.

[153] Huang, Z; Negishi, E.-i. *Org. Lett.* **2006**, *8*, 3675 – 3678.

[154] a) Saito, A.; Ito, H.; Taguchi, T. *Org. Lett.* **2002**, *4*, 4619 – 4621; b) Saito, A.; Yanai, H.; Taguchi, T. *Tetrahedron Lett.* **2004**, *45*, 9439 – 9442; c) Saito, A.; Yanai, H.; Taguchi, T. *Tetrahedron* **2004**, *60*, 12239 – 12247.

[155] Ishikawa, T.; Senzaki, M.; Kadoya, R.; Morimoto, T.; Miyake, N.; Izawa, M.; Saito, S.; Kobayashi, H. *J. Am. Chem. Soc.* **2001**, *123*, 4607 – 4608.

[156] Reppe, W. et al. *Liebigs Ann. Chem.* **1955**, *596*, 38 – 79.

[157] Barluenga, J.; González, J. M.; Rodríguez, M. A.; Campos, P. J.; Asensio, G. *Synthesis*, **1987**, 661 – 662.

[158] Ondruš, V.; Orsága, M.; Fišera, L.; Prónayová, N. *Tetrahedron* **1999**, *55*, 10425 – 10436.

[159] Khanapure, S.P.; Najafi, N.; Manna, S.; Yang, J.; Rokach J. *J.Org.Chem.* **1995**, *23*, 7548-7551.

[160] Kessinger, R.; Thilgen, C.; Mordasini, T.; Diederich, F. *Helv. Chim. Acta* **2000**, *83*, 3069-3096

Graphical Abstract

ABSTRACT

The classical drugs used for the treatment of common bacterial diseases are often characterised by serious side effects and high-toxicity profiles. Additionally, in the last decade a rapid development of multidrug-resistant strains of bacterial pathogens has been observed. Consequently, there is an urgent need to discover new structural classes of antibacterial compounds, and to develop agents which are able to replace (or be associated with) the drugs which are currently in use. In 1998, in a research program directed towards the isolation and characterisation of such substances, researchers at the University of Göttingen isolated from a stem of *Streptomyces* a novel antibiotic, named branimycin.

To access the complex structure of branimycin by total synthesis, the molecule was divided retrosynthetically into two fragments: the *cis*-octalin core and the polyketide side-chain.

In this thesis, 3 approaches are described to access the *cis*-octalin core fragment through substrate-controlled intramolecular (IMDA) or transannular (TADA) Diels-Alder reactions. The best results were obtained when the diastereocontrol of the TADA reaction was exerted by a chiral lactol tether, which translated its chirality to the *cis*-octalin core. The geometrically defined *Z,E,Z,E*-tetraene, necessary for the diastereoselective Diels-Alder reaction, was constructed from a chiral lactol, itself derived from L-ascorbic acid. The tetraene was then constructed in a Z-selective Julia-olefination, a novel oxidation-olefination sequence followed by an intramolecular Stille-coupling. After the stereoselective Diels-Alder reaction, the product could then further be successfully converted to a *cis*-octalin epoxide, carrying all necessary functionality.

Furthermore, the side-chain fragment could be constructed in a convergent manner by the reaction of a differentially protected glyceraldehyde derivative with a chiral allenyl silane. The required *syn,syn* diastereomer was obtained as the major product through an *anti*-Felkin-Anh transition state. Further conversion of this stereotriad led to formation of a orthogonally protected, metallated side-chain fragment which was successfully added to a (branimycin-) test substrate.

ZUSAMMENFASSUNG

Die klassischen Wirkstoffe zur Behandlung von bakteriellen Erkrankungen sind oft durch ernste Nebenwirkungen und/oder Toxizitätsprobleme gekennzeichnet. Zusätzlich dazu haben im letzten Jahrzehnt einige Stämme pathogener Keime Resistenzen gegen mehrere der gängigen Antiobiotika entwickelt. Es ist daher dringend erforderlich, mithilfe der Identifizierung strukturell neuartiger Antibiotikaklassen und deren Entwicklung zu Medikamenten, die augenblicklich verwendeten Wirkstoffe zu ersetzen bzw. Kombinationen mit ihnen zu ermöglichen. 1998 wurde an der Universität Göttingen, im Rahmen eines Forschungsprogramms zur Isolierung und Identifizierung neuer Antibiotika, aus einem Stamm von Streptomyceten ein neuartiges Antibiotikum isoliert und Branimycin genannt.

Der totalsynthetische Zugang zur komplexen Struktur von Branimycin erfolgt durch die retrosynthetische Zerlegung in zwei Fragmente, das *cis*-Oktalin Kernfragment und die polyketidische Seitenkette.

In dieser Dissertation werden 3 verschiedene Zugänge zur Herstellung des Kernfragments unter Substratkontrolle in einer intramolekularen (IMDA) bzw. einer transannularen Diels-Alder (TADA) Reaktion beschrieben. Die besten Ergebnisse wurden mit einer diastereoselektiven TADA Reaktion erhalten – die chirale Information wurde hierbei von einem *trans*-Lactol auf den *cis*-Oktalin Kern übertragen. Das für die diastereoselektive Diels-Alder Reaktion notwendige geometrisch einheitliche Z,E,Z,E-Tetraen wurde ausgehend von einem aus L-Ascorbinsäure abgeleiteten chiralen Lactol aufgebaut. Das Tetraen wurde dann mittels einer Z-selektiven Julia-Olefinierung und einer neuartigen Oxidations-Olefinierungssequenz, gefolgt von einer intramolekularen Stille-Kopplung, hergestellt. Nach der stereoselektiven Diels-Alder Reaktion konnte das Produkt zu einem vollständig funktionalisierten *cis*-Oktalinepoxid umgesetzt werden.

Weiters konnte das Seitenkettenfragment auf konvergente Weise durch die Reaktion eines differentiell geschützten, Glyceraldehyd-derivats mit einem axial-chiralen Allenylsilan hergestellt werden. Das benötigte *syn,syn*-Diastereomer wurde als Hauptprodukt über einen *anti*-Felkin-Anh Übergangszustand erhalten. Diese Stereotriade konnte in die orthogonal geschützte, metallierte Seitenkette überführt werden, die auch erfolgreich an ein (Branimycin-)Testsubstrat addiert werden konnte.

I want morebooks!

Buy your books fast and straightforward online - at one of the world's fastest growing online book stores! Environmentally sound due to Print-on-Demand technologies.

Buy your books online at
www.get-morebooks.com

Kaufen Sie Ihre Bücher schnell und unkompliziert online – auf einer der am schnellsten wachsenden Buchhandelsplattformen weltweit!
Dank Print-On-Demand umwelt- und ressourcenschonend produziert.

Bücher schneller online kaufen
www.morebooks.de

SIA OmniScriptum Publishing
Brivibas gatve 1 97
LV-103 9 Riga, Latvia
Telefax: +371 68620455

info@omniscriptum.com
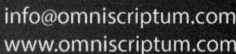

Printed by Books on Demand GmbH, Norderstedt / Germany